Smart Building Systems for Architects, Owners, and Builders

Smart Building Systems for Architects, Owners, and Builders

James Sinopoli

AMSTERDAM • BOSTON • HEIDELBERG • LONDON
NEW YORK • OXFORD • PARIS • SAN DIEGO
SAN FRANCISCO • SINGAPORE • SYDNEY • TOKYO

Butterworth-Heinemann is an imprint of Elsevier

Butterworth-Heinemann is an imprint of Elsevier
30 Corporate Drive, Suite 400
Burlington, MA 01803, USA

The Boulevard, Langford Lane
Kidlington, Oxford, OX5 1GB, UK

Notices
Knowledge and best practice in this field are constantly changing. As new research and experience
broaden our understanding, changes in research methods, professional practices, or medical
treatment may become necessary.

Practitioners and researchers must always rely on their own experience and knowledge in
evaluating and using any information, methods, compounds, or experiments described herein. In
using such information or methods they should be mindful of their own safety and the safety of
others, including parties for whom they have a professional responsibility.

To the fullest extent of the law, neither the Publisher nor the authors, contributors, or editors,
assume any liability for any injury and/or damage to persons or property as a matter of products
liability, negligence or otherwise, or from any use or operation of any methods, products,
instructions, or ideas contained in the material herein.

Library of Congress Cataloging-in-Publication Data
Application submitted.

ISBN: 978-1-85617-653-8

British Library Cataloguing-in-Publication Data
A catalogue record for this book is available from the British Library.

For information on all Butterworth–Heinemann publications
visit our Web site at *www.elsevierdirect.com*

Working together to grow
libraries in developing countries

www.elsevier.com | www.bookaid.org | www.sabre.org

ELSEVIER   BOOK AID
             International   Sabre Foundation

Transferred to Digital Printing in 2011

Contents

Preface

Technology has always influenced the buildings we build, and always will. Twenty-five to 30 years ago, however, the amount of technology in a building was minimal. It consisted of the public telecommunications utility installing its services in a building; a mechanical contractor installing a pneumatic control system for the heating, cooling, and ventilation system; and maybe a word-processing system. Although we have come a long way since those days, we are still in a very early stage of fully deploying and integrating technology systems into buildings.

In due course buildings will become full of technology. Walls and ceilings will be embedded with sensors, and every aspect of a building's performance and use will be metered and measured. Software tools will be used to automatically optimize building systems without human intervention; real-time information about the building that is relevant to their particular needs will be provided to occupants and building management. Buildings will be fully interactive with the power grid, and geospatial location systems will be deployed for every building asset.

I wrote this book as a step toward eventually fulfilling that vision. It is meant as a guide to understanding the many aspects needed to deploy integrated technology systems into buildings and to provide straightforward information on smart buildings for architects, engineers, facility managers, developers, contractors, and design consultants. What's here reflects my personal experience and research, and information gained from listening to and learning from many colleagues.

Smart buildings can be many things, but simply defined: smart buildings use building technology systems to enable services and the operation of a building for the betterment of its occupants and management. The drivers for smart buildings are the positive financial effects of integrated systems, energy conservation, greater systems functionality, and the continuing evolution of technology. The headwind to smart buildings is the inertia of people to move beyond the legacies of building design, construction, and operation. Such processes as Building Information Modeling as well as the movement to energy-efficient and sustainable buildings are beginning to change that, however.

Specialists in certain technologies may find the coverage of some of the systems in this book to be elementary but can gain knowledge of other technology systems they may be less familiar with. To deal with a smart building one has to be somewhat of a generalist, understanding the synergy principal: "the whole is greater than the sum of its parts." It also helps to know something about each of a building's technology systems, as well as the processes needed to design, construct, and operate a building.

Acknowledgments

I want to thank and acknowledge several people whose input and influence helped shape this book. I'm fortunate to work every day with three exceptional individuals: Neil Gifford, who is simply one of the best building controls and system integration consultants on the planet; Gina Elliott, an energetic woman with extensive experience in business, technology, and integrated systems; and Andres Szmulewicz, a quiet, methodical, and extremely competent man who I've teamed with for years. I am also grateful to Christopher Rendall, a fine young engineer from the University of Texas who helped with research. Last, but not least, I need to thank my wife Kate for her endless patience and counsel.

What Is a Smart Building?

Brief History

Smart buildings, or at least discussion of the concept, originated in the early 1980s. In 1984, for instance, a *New York Times* article described real estate developers creating "a new generation of buildings that almost think for themselves ... called intelligent buildings." Such a building was defined as "a marriage of two technologies—old-fashioned building management and telecommunications."

In the early 1980s, several major technology trends were under way. One was that the U.S. telecommunications industry was undergoing deregulation and new companies, products, services and innovations entered the telecom marketplace. The second major trend, which at the time seemed somewhat separate and unrelated, was the creation and emergence of the personal computer industry. This era also spawned the first real connection between real estate developers and technology. The newly unregulated telecommunications industry presented an opportunity for building owners to resell services within their facilities and add value to their business. This new business model was known as "shared tenant services."

Under shared tenant services, the building owner procured a large telecommunications system for the entire building and leased telecommunication services to individual tenants. Major real estate developers offered such shared services but eventually abandoned such arrangements due to inadequate profitability and lack of knowledge and skills in telecommunications. It was, however, one of the first times that building owners thought about and acted on the idea of major technology systems in buildings.

In the next decade or so, there were some modest technological advancements in buildings, including structured cabling systems, audio visual systems, building automation controllers with direct digital control (DDC), conditioned space for network equipment, access control systems, and video surveillance, among others. Yet guidelines for building construction documents released in 1994, the Construction Specifications Institute's *MasterFormat*, had 16 divisions, barely mentioning technology. Many times engineers and designers used a "Division 17" for the specification of technology-related systems.

Division 17 was not a formal specification division but was used for materials and equipment not included in the other 16 divisions. During that time period a traditional mind-set prevailed among most building designers in which technology was an afterthought rather than integral to the building design. The latest revision of the *MasterFormat* in 2004 was an improvement, but still lags in terms of technological advances in buildings. It is evident that technology is advancing more rapidly and probably progressing through several life cycles during the time it takes to revise the construction specification format guidelines.

Smart buildings are not just about installing and operating technology or technology advancements. Technology and the systems in buildings are simply enablers, a means to an end. The technology allows us to operate the building more efficiently; to construct the buildings in a more efficient way, to provide productive and healthy spaces for the occupants and visitors, to provide a safe environment, to provide an energy-efficient and sustainable environment, and to differentiate and improve the marketability of the building.

What Is a Smart Building?

A smart building involves the installation and use of advanced and integrated building technology systems. These systems include building automation, life safety, telecommunications, user systems, and facility management systems. Smart buildings recognize and reflect the technological advancements and convergence of building systems, the common elements of the systems and the additional functionality that integrated systems provide. Smart buildings provide actionable information about a building or space within a building to allow the building owner or occupant to manage the building or space.

Smart buildings provide the most cost effective approach to the design and the deployment of building technology systems. The traditional way to design and construct a building is to design, install, and operate each system separately (Fig. 1.1).

The smart building takes a different approach to designing the systems. Essentially, one designer designs or coordinates the design of all the building

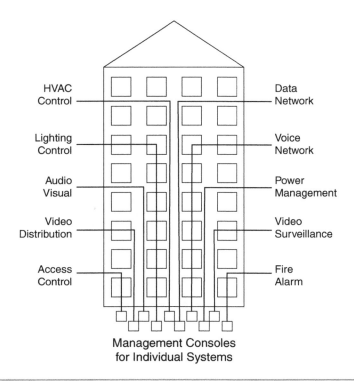

Figure 1.1 Multiple proprietary building systems.

technology systems into a unified and consistent construction document. The construction document specifies each system and addresses the common system elements or integration foundation for the systems. These include cabling, cable pathways, equipment rooms, system databases, and communications protocols between devices. The one consolidated design is then installed by a contractor, referred to as a Technology Contractor or as a Master System Integrator.

This process reduces the inefficiencies in the design and construction process saving time and money. During the operation of the building, the building technology systems are integrated horizontally among all subsystems as well as vertically—that is subsystems to facility management systems to business systems—allowing information and data about the building's operation to be used by multiple individuals occupying and managing the building (Fig. 1.2).

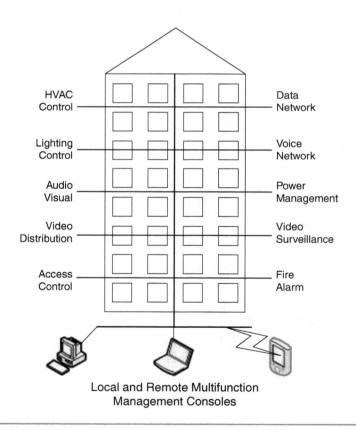

Figure 1.2 Integrated building systems.

Smart buildings are also a critical component regarding energy usage and sustainability of buildings and the smart electrical grid. The building automation systems, such as HVAC control, lighting control, power management, and metering play a major role in determining the operational energy efficiency of a building. The smart electrical grid is dependent on smart buildings.

The driving forces for smart buildings are economics, energy, and technology. Smart buildings leverage mainstream information technology infrastructure and take advantage of existing and emerging technology. For developers and owners, smart buildings increase the value of a property. For property and facility managers, smart buildings provide more effective subsystems and more efficient management options, such as the consolidation of system management. For architects, engineers, and construction contractors, it means combining portions of the design and construction with the resulting savings and efficiencies in project management and project scheduling.

Chapter 2

The Foundations of a Smart Building

Overview

All technology systems in a building are networks consisting of end devices that communicate with control devices or servers to monitor, manage, or provide services to the end devices. Communications between the devices occur via a set of rules or protocols. Connectivity between devices on the network is either through cable or a wireless transmitter/receiver. The network typically has a system administration workstation or PC that provides a management and reporting function.

Doi:10.1016/B978-1-85617-653-8.00002-8

In many systems, databases are associated with the network such as security access credentials and lighting schedules. Recognition of these network commonalities together with the utilization of typical information technology infrastructure comprises the core of smart buildings and the integration foundations of building technology systems.

Smart buildings are built on open and standard communications networks which make the following characteristics possible: (1) inter-application communication; (2) efficiencies and cost savings in materials, labor, and equipment; and (3) interoperable systems from different manufacturers.

The Framework for Referencing Integration

Building system integration takes place at physical, network and application levels. Integrated systems share resources. This sharing of resources underpins the financial metrics and improved functionality of integrated systems.

System integration involves bringing the building systems together both physically and functionally. The physical dimension obviously refers to the cabling, space, cable pathways, power, environmental controls, and infrastructure support. It also touches on common use of open protocols by the systems. The functional dimension refers to an interoperational capability, this means integrated systems provide functionality that cannot be provided by any single system, the whole is greater than the sum of the parts.

There is a key differentiation between integrated and interfaced systems. Interfaced systems are essentially standalone systems that share data, but continue to function as standalone systems. Integrated systems strive for a single database, a meta-database, thus reducing the cost and support for synchronizing separate databases.

At the forefront of the evolution to open network standards is the International Standards Organization's (ISO) development of the Open System Interconnection (OSI) model. The OSI model presents seven layers of network architecture (the flow of information within an open communications network), with each layer defined for a different portion of the communications link across the network. This model has withstood the test of time and its framework and derivatives should serve as the reference point for network integration (Fig. 2.1).

The model is straightforward. A network device or administrator creates and initiates the transmission of data at the top layer (the application layer), which moves from the highest layer to the lowest layer (physical layer) to communicate the data to another network device or user. At the receiving device the data travel from the lowest layer to the highest layer to complete the communication. When

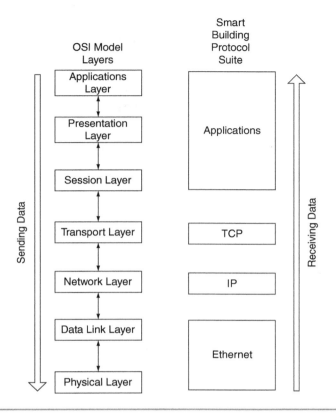

Figure 2.1 Network model layers.

the data packet is initially sent each layer takes the data of the preceding layers and adds its own information or header to the data. Basically, each layer puts its own "envelope" around the preceding "envelope." On the receiving end each layer removes its information or "envelope" from the data packet. The layers or groups of layers are described in the following sections.

Physical Layer

This layer defines the electrical (or in the case of fiber optic cables, light) communications across a network link or channel. This layer guarantees that bits of data transmitted by a device on the network are accurately received by another device on the network. The physical layer initiates, maintains, and terminates the physical connection between network devices. It defines the mechanical and the electrical characteristics of the physical interface, including connectors, network interface cards, and voltage and transmission distances.

The network protocol RS-232, once heavily utilized in telecommunications and data networks, is defined solely by the physical layer. Other network protocols, such as Ethernet, are defined by the physical layer as well as some of the entire next layer in the stack—the data link layer.

DATA LINK LAYER

The data link layer takes the data bits and "frames," and creates packets of the data to guarantee reliable transmission. This layer adds source and destination addresses to the data stream as well as information to detect and control transmission errors. The data link layer has two sublayers. One is the logical link control (LLC) sublayer, which essentially maintains the communications link between two devices on the network. The other is the media access control (MAC) sublayer which manages the transmission of data between two devices. The network card on a PC has a MAC address, essentially a unique address for every device on a local area network.

The details of the data link layer can be specified differently and are reflected in various network types (Ethernet, token ring, etc.). Each network type has its own method of addressing, error detection, control of network flow, and so forth.

NETWORK LAYER

The network layer routes data packets through the network. It deals with network addressing and determines the best path to send a packet from one network device to another. The Internet Protocol (IP) is the best example of a network layer implementation.

TRANSPORT LAYER

The transport layer is responsible for reliable transport of the data. At times it may break upper-layer data packets into smaller packets and then sequence their transmission. The Transmission Control Protocol (TCP), one of the major transport protocols, is typically used with the best-known network layer protocol, IP, and is referred to as TCP/IP.

SESSION, PRESENTATION, AND APPLICATION LAYERS

Many times the session, presentation, and application layers are considered as one layer. The session and presentation layers manage dialogue between end-user applications, then format and deliver the data to the application layers. The Session Layer establishes, manages, and terminates the connection between the local and remote application. The Presentation Layer establishes a data framework between the Application Layer entities. It translates the data representation in an application to a network format and vice versa. The application layer is the layer the end user directly interacts with.

The discussion of system integration should be framed using the ISO model, focusing on the physical, data, network, and application layers. Doing so adds clarity and understanding to both industry and client discussions.

Structured Cabling Systems

Over the last 20 years there has been considerable movement away from proprietary telecommunications cable to generic cable that is independent of individual vendors or specific uses. There are several authoritative standards for generic structured cable infrastructure for telecommunications systems used throughout the world.

The first cable standard created was the ANSI/TIA/EIA-568 A and B Commercial Building Telecommunications Cabling Standard published by the American National Standards Institute (ANSI), the Telecommunications Industry Association (TIA), and the Electronic Industry Association (EIA). Canada has a standard similar to ANSI/TIA/EIA-568 referred to as CSA T529 (Canadian Standards for Telecommunications Wiring Systems).

There is also an international standard referred to as ISO/IEC 11801 which is used in Asia, Europe, and Africa. The international standard was authored by the International Organization for Standardization (ISO) and the International Engineering Consortium (IEC) in 1995. The European Union has a standard similar to ISO/IEC 11801 called EN 50173 and was published by the European Committee for Electrotechnical Standardization (Cenelec).

Australia and New Zealand have published AS/NZS 3080, also a standard for generic cabling infrastructure. Although there are some differences between the ANSI/TIA/EIA-568 standard and the other standards, the ANSI/TIA/EIA-568 standard has served as a basis for deriving the other standards (Fig. 2.2).

In 2002 the ANSI/TIA/EIA standards organizations also published ANSI/TIA/EIA-862, Building Automation Systems Cabling Standard for Commercial Buildings, which is essentially documentation of similar standards for building

ANSI/TIA/EIA-862-2002
Approved: April 11, 2002

TIA/EIA
STANDARD

Building Automation Systems Cabling Standard for Commercial Buildings

TIA/EIA-862

APRIL 2002

TELECOMMUNICATIONS INDUSTRY ASSOCIATION

Figure 2.2 Cable standard for building automation systems.

automation systems (BAS). These two standards, ANSI/TIA/EIA-568 and ANSI/TIA/EIA-862, are applicable to all of the smart building technology systems, with the exception of fire alarm systems. Both standards allow for an open cable infrastructure that is independent of specific products and vendors.

The primary basis for both standards is the use of unshielded twisted-pair (UTP) copper and fiber optic cable. Both standards have similar design guidelines and parameters for reliability, capacity, and compatibility. However, there are two

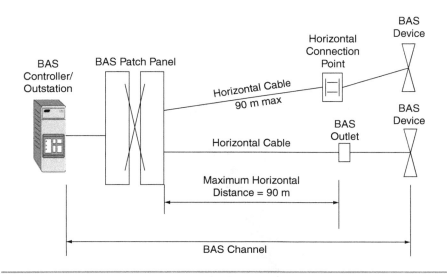

Figure 2.3   Cabling configurations for BAS using structured cable.

minor differences in the BAS standard: (1) a "horizontal connection point," which is similar to telecommunications "consolidation points" or zone cabling, and (2) a "coverage area" as opposed to telecommunications "work area" (Fig. 2.3).

Despite minor differences, these two standards allow for a single structured cabling system for a building. Until the recent standard on BAS cabling was developed, building automation systems were cabled separately using different cable types and cable pathways.

In addition, these systems have traditionally used hardwired connections of cables from the BAS equipment to the devices unlike the ubiquitous twisted-pair RJ-45 connections of the telecommunications network world. This is currently evolving to increased use of standard cable infrastructure based on unshielded twisted-pair copper and fiber optic cables.

It should be noted that twisted-pair copper and fiber optic cables are used extensively in these networks but not all end devices warrant the use of such cables. For example, twisted-pair cables may be used to connect a PC to a network but would not be used to connect a mouse or keyboard to the PC. Similar situations exist with other end devices in other smart building systems.

TWISTED-PAIR COPPER CABLE

The core of a twisted-pair copper cable (Fig. 2.4) is made up of two insulated copper cables twisted together into a "pair." Four pairs are jacketed together

Figure 2.4   Twisted pair copper cable.

for a standard four-pair copper cable. The wires are relatively thin (between 22 and 24 gauge). The cables are twisted into a pair to reduce crosstalk ("coupling" of the pairs or interference from an adjoining cable) as well as interference from electrical and mechanical sources. Each pair has a different number of twists relative to the other pairs in the cable to further reduce crosstalk. This construction is usually referred to as unshielded twisted pair.

The twisted pairs can also be "shielded," creating shielded twisted-pair cable. While the use of shielded twisted pair is popular in some countries and has some technical advantages, it is more labor intensive during installation because of the grounding required and is typically not used in a structured cable infrastructure.

Unshielded twisted-pair cables are relatively inexpensive and there are a large number of technicians qualified to install the cable. The cable standards guarantee performance of the cable over 90 meters (295 feet). Categories of unshielded twisted pairs are based on the bandwidth, or information-carrying capacity, of the cable.

The most recent categories of unshielded twisted-pair are Category 5e and Category 6. Category 5e cable is specified up to a bandwidth of 100 MHz (Hz is a unit of frequency equal to one cycle per second). The Category 6 standards set requirements up to 250 MHz. These standards not only apply to the cable but also to all connector and cable termination devices such as the cable jacks, patch cords, and the patch panel. Many manufacturers test and/ or manufacture cable beyond the standards to differentiate their product as going beyond the standards.

FIBER OPTIC CABLE

Fiber optic cables use strands of glass to propagate light. The light pulses transport communication signals between devices. At the center of the fiber optic strand is a small inner core that carries the propagated light. Surrounding the core is the outer cladding. Both the core and the cladding are glass but have different "refractive indexes" which essentially means that light travels at different speeds through the materials. The result is that light pulses produced from lasers or LEDs at one end of a fiber optic cable are sent through the fiber optic core and are reflected back to the core when the light hits the fiber optic cladding, thus keeping the light within the center core.

Fiber optic cables with small inner cores (10 microns or less) have only one path for the light and are referred to as single-mode fiber. Fiber optic cables with slightly larger cores (50 and 62.5 microns) have multiple paths for the light and are referred to as multimode fiber. The cladding of both types of fiber is 125 microns (Fig. 2.5). For comparison, human hair is generally in

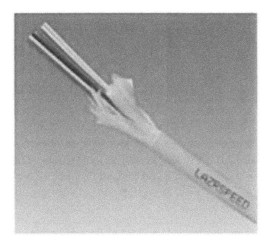

Figure 2.5 Fiber optic cable.

the range of 40 to 120 microns (a micron is one millionth of a meter, or about 0.00004 inch).

As the light pulse transverses the cable it loses power. This loss, called attenuation, is measured in decibels. The attenuation of fiber optic cables is dependent on the wavelength. Multimode fiber optic cables operating at low wavelengths may have attenuation less than 3.5 decibels per kilometer and less than 1.5 decibels per kilometer at higher wavelengths.

Single-mode fiber optic cables have superior performance with maximum attenuation of 0.5 to 1 decibel per kilometer and can be used anywhere in a network but are generally used for longer distances. Multimode fiber optic cables are utilized for shorter distances (less than a mile) and are found generally within buildings. Beyond the high bandwidth capacity of fiber optic cables major benefits include immunity from electromechanical interference, immunity from radio frequency interference, and use over longer distances.

COAXIAL CABLE

For many years coaxial cable was the cable of choice with extensive use in video distribution systems, early Ethernet network installations, and connectivity to many large mainframe computer systems. Coaxial cable is essentially a wave guide transmitting radio and television frequencies down the cable and is immune to electromechanical interference. It continues to be used for some video transmission. However, with technical advances in IP video and the use of "baluns" (which allow video signals to be transmitted over unshielded twisted-pair cable) the use of coaxial cable is decreasing. Its use in a smart building is generally minimal.

Wireless

Wireless connectivity is just a substitute for cabled connectivity. Wireless does not and technically cannot provide the theoretical bandwidth of a physical cable connection. However, wireless can provide mobility and is an excellent option for connectivity in older buildings where pathways for cable may not be available. The wireless technologies probably most useful for smart buildings technology systems include Wi-Fi and an emerging technology, Zigbee.

WI-FI

Wireless Fidelity (Wi-Fi) systems basically replace cabled Ethernet connections with a wireless device. Current Wi-Fi systems operate in two unlicensed radio frequencies, 2.4 GHz and 5 GHz. The Institute for Electrical and Electronics Engineers (IEEE) has set three standards for Ethernet communications via these frequencies, which are commonly referred to as IEEE 802.11a, operating in the 5-GHz frequency, and IEEE 802.11b and 802.11g, operating in the 2.4-GHz frequency. These standards can provide "optimal" throughput of 11 Mbps (a measure of bandwidth, megabits per second) and 54 Mbps. Another standard, IEEE 802.11n, has been proposed and is expected to be approved in late 2009. IEEE 802.11n will have a throughput of 110 Mbps (Fig. 2.6).

The user's distance from the antenna, the utilization of the same unlicensed frequencies by other devices, obstacles inside buildings, and building structures that interfere with the radio signals all affect the communications bandwidth received from the Wi-Fi antenna.

Typical coverage areas indoors for omnidirectional Wi-Fi antenna are 100 to 300 feet. Each wireless access point (WAP) or gateway can generally serve 10 to 20 users depending on their applications. This technology and wireless "hot spots" are now common in public buildings, airports, businesses, hotels, restaurants and homes. The marketplace for Wi-Fi equipment is moving toward "wire-line–class" security, high-performance, reliability, and enterprise-scale manageability of systems.

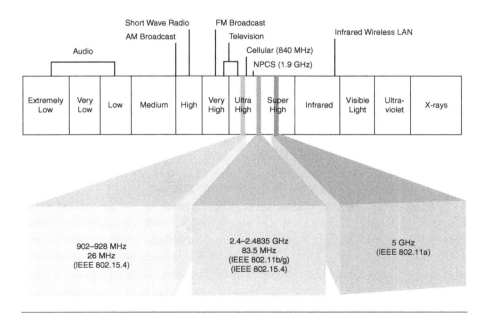

Figure 2.6   Radio frequencies for Wi-Fi.

WI-FI MESH NETWORKING

In a wireless mesh network all nodes or WAPs are interconnected wirelessly instead of using cable to connect wireless antennas. Typically these "hot spots" are created using omnidirectional Wi-Fi IEEE 802.11b/g antennas. Directional antennas using Wi-Fi IEEE 802.11a are then used to connect the hot spots or nodes and create the mesh. The 802.11a standard is used in the mesh backbone because of its performance (54 Mbps) as well as different and noninterfering radio frequency (IEEE 802.11a uses 5 GHz, while 802.11b/g antennas use 2.4 GHz).

The nodes on the mesh network automatically learn about one another and self-configure network traffic through numerous network path configurations. The result is extended coverage of a Wi-Fi network with the ability to route traffic around congestion and around obstacles and interference. Mesh networks provide redundancy and "robustness" because of their capability to balance network traffic.

Wireless mesh networks create latency that can affect applications such as voice over IP. They can also be noisy, possibly creating transmission errors, resulting in retransmissions and reductions of available bandwidth. While current implementations are proprietary, a standard for Wi-Fi mesh networks, called 802.11s was finalized and released in 2007.

ZIGBEE

Zigbee is a wireless technology standard (IEEE 802.15.4) that provides for low data-rate networks. It uses unlicensed frequencies (900 MHz and 2.4 GHz), which are also available for cordless phones, Wi-Fi, and other devices. The standard is aimed at control devices in residential, commercial and industrial buildings. It is particularly useful for sensors and control devices in building automation systems in a smart building where very small amounts of information or data are being transmitted. This includes on/off switches, open/closed devices, thermostats, and motor controls (Fig. 2.7).

The maximum speed of Zigbee devices varies between 192 and 250 Kbps (a measure of bandwidth, kilobits per second). The maximum distance varies between 20 and 50 meters. Zigbee has several advantages:

■ Low power usage as the devices require only two AAA batteries
■ Wide support from more than 100 companies supporting the standard (companies such as Motorola, Honeywell, Samsung, and Mitsubishi)

IEEE Std 802.15.4™-2003

802.15.4™

IEEE Standard for
    Information technology—
Telecommunications and information
    exchange between systems—
Local and metropolitan area networks—
Specific requirements

Part 15.4: Wireless Medium Access Control (MAC) and Physical Layer (PHY) Specifications for Low-Rate Wireless Personal Area Networks (LR-WPANs)

IEEE Computer Society

Sponsored by the
LAN/MAN Standards Committee

◈IEEE

Published by
The Institute of Electrical and Electronics Engineers, Inc.
3 Park Avenue, New York, NY 10016-5997, USA

1 October 2003

Figure 2.7   Zigbee standard.

- Mesh technology that allows Zigbee, like Wi-Fi, to be configured in several topologies, including a mesh topology allowing multiple transmission paths between the device and the recipient
- System scalability where thousands of Zigbee devices can deployed in a building

Communications Protocols

Protocols are the set of rules that define a format to communicate data between network devices and other networks. Protocols may address the rate or speed of the communications transmission, the mode, the error detection and the encoding and decoding of data. Protocols can be implemented by hardware, software, or a combination of the two.

Many years ago data processing was done solely on mainframe computers or other centralized computers. These systems had proprietary communications between the user's "dumb" terminal and the mainframe equipment, both of which were manufactured by the same company. For example, one could not take a user's terminal manufactured by IBM Corporation and use it on a system made by Digital Equipment Corporation.

Today's data networks use PCs and devices from multiple manufacturers. Different PCs can be networked, not only because they have processing power and storage that the "dumb" terminals did not have but more importantly, because they communicate through a common, open communications protocol. The technology systems in a smart building have adopted and are migrating to the common network protocols used in data networks (Fig. 2.8).

Figure 2.8 Ethernet connection.

ETHERNET

The general technology trend is for everything to become part of a node on a cabled or wireless Ethernet network and have an IP address. This is the digital convergence trend we have already witnessed and will continue to see for years to come.

In the data telecommunications arena IP and Ethernet have been standards for years. Voice communications, traditionally modeled after legacy monolithic mainframe computers, are quickly moving to the data telecommunications world via voice over IP (VoIP). Video is also moving to digital communication protocols. Building automation systems have specific industry protocols (BACnet, LonTalk and others), but they too are moving to convert or interface the protocols to the universal, dominant IP protocol.

Life safety systems still lag somewhat, video surveillance systems have moved to IP, and access control is moving in that direction as well. Even legacy systems with proprietary standards can have protocols converted or translated to the standard IP protocols.

XML AND SOAP

EXtensible Markup Language (XML) is a structure for storing data in a plain text format and transporting the data. XML is important for smart buildings because it is independent of hardware and software and is the most common method for transmitting data between system applications. It normalizes and standardizes data and facilitates the integration of systems and system databases that may contain data in incompatible formats.

For example, you can take security credentials data from a Structured Query Language (SQL) database for access control, convert the data to XML, then share the data with a human resources or student record application; this would allow security credentials to be created or deleted as employees or students move in and out of the organization without the need to coordinate and duplicate the actions in both databases. In addition, a very large number of international programmers and software vendors have adopted XML.

The XML data are "wrapped" in tags that are specific to the application. Tags are created by a programmer to send, receive, or display the data. Unlike HTML, which formats and displays data, XML has no predefined tags.

Whereas XML formats data, the Simple Object Access Protocol (SOAP) is used to communicate that data between applications. SOAP is a communications protocol based on XML that communicates via the Internet. Like XML, SOAP is independent of hardware platforms, programming languages, and operating systems, and is important in facilitating the exchange of information between applications and systems.

BACNET

BACnet is an acronym for building automation and control networks. An international data communication protocol, BACnet was first published in 1996, and was developed and is maintained by the American Society of Heating, Refrigerating and Air-Conditioning Engineers (ASHRAE).

BACnet provides a standard for representing the functions and operations of building automation and control devices. For example, the protocol covers how to request a value from a humidity sensor or send a pump status alarm.

For each type of building automation and control device, a standard software object is created that contains the identifier and properties of such a device reflecting the functions and operation of the device. Some of these properties may be inherent or required properties of the device, while other properties may be optional features of the device. The essence of BACnet is to move away from proprietary communications to similar devices by different manufacturers and treat communications and control of like-devices in a standard common way. This approach and structure allow BACnet to be used in HVAC, lighting systems, fire alarm systems, and other building automation systems.

In addition to standardization of the device objects, BACnet also defines the message types between a server and a client. These messages are called "service requests," and the BACnet standard defines 35 message types that are divided into five groups or classes. The latter involve messages for the following:

1. Accessing and manipulating the properties of the objects
2. Alarms and events
3. File uploading and downloading
4. Managing the operation of remote devices
5. Virtual terminal functions

BACnet can communicate over several types of networks, including Ethernet, ARCNET, MS/TP (master-slave/token-passing), and PTP (point-to-point) for use over phone lines or hardwired EIA-232 connections.

ARCNET is an archaic network topology and PTP is not widely deployed. ASHRAE has recognized the importance of the IP protocol in communications. ASHRAE initially created a method called "Annex H.3," allowing for the BACnet protocol to be "tunneled" through an IP network (i.e., basically packaging the BACnet data into an IP message). However this required each subnetwork to have a translator that could assemble and disassemble the data in the IP packet. The shortcoming of Annex H.3 was addressed with Annex J BACnet/IP. BACnet/IP transmits BACnet messages in native IP format, providing new flexibility and scalability to BACnet.

LonWorks

LonWorks is often referred to as a communications protocol for control networks, but because it bundles a communications protocol with a dedicated microprocessor and media transceivers, it more closely resembles a networking platform. LonWorks was created by the Echelon Corporation. In 1999, the communications protocol (then known as LonTalk) was submitted and accepted as a standard for control networking (ANSI/CEA-709.1-B).

In 2009, LonWorks became an international standard, ISO/IEC 14908. Whereas LonTalk addresses the issue of how devices communicate, LonWorks defines the content and structure of the information that is communicated. The protocol is primarily focused on building and home automation, but is also used in transportation and industrial automation.

The standard calls for two primary physical-layer signaling technologies; twisted-pair cable and a power line carrier, although LonWorks can also use radio frequency (RF), infrared (IR), coaxial cable and fiber optic cable. The LonWorks platform uses an affiliated IP tunneling standard—ANSI/CEA-852—in use by a number of manufacturers to connect the devices on LonWorks-based networks to IP networks and applications. Many LonWorks networks are deployed with some IP network integration.

LonWorks is primarily implemented using a device that contains an Echelon-designed 8-bit processor, called the "Neuron chip." Similar to BACnet objects, LonWorks standardizes the functions and information of devices. Each standard is known as standard network variable types (SNVTs).

Modbus

Modbus is a communications protocol published by Modicon in 1979. At that time Modbus primarily focused on communication to programmable logic controllers (PLCs) manufactured by Modicon and used in industrial automation. Modicon is currently a company owned by Schneider Electric and in 2004 the Modbus standard was transferred to a nonprofit organization,

Modbus-IDA, whose members are primarily users and suppliers in the automation industry.

Modbus is an application-layer messaging protocol for client–server communication between devices connected on different types of buses or networks. It can be implemented over an Ethernet network as an asynchronous serial transmission such as RS-232 or RS-485, or as a high-speed, token-passing network call Modbus Plus. For Ethernet and Modbus Plus the message created by the Modbus protocol is "tunneled" or imbedded into the frame or packet structure that is used on an Ethernet or token-passing network.

The most common implementation of Modbus uses the serial RS-485 physical layer with either Modbus RTU (a binary representation of the data) or Modbus ASCII (human readable). The Ethernet implementation option uses Modbus/TCP.

Modbus, like other communication protocols, defines a message structure and format for message fields; how a controller requests access to another device, how to respond to requests, how errors will be detected and reported, how to identify devices, how to recognize a message to a device, and so on.

Modbus versions have different functionality. For example, the basic Modbus protocol is a master–slave arrangement that does not provide for a "slave" device to report to the master unless it is polled by the master. In the Modbus implementation over Ethernet, Modbus/TP devices can report to a master. Typical Modbus implementations are limited to 247 devices, although in a Modbus/TP implementation, no such limit exists.

Modbus is a simple yet effective protocol. Typical problems that the designers have to overcome include high latency and timing problems.

OPC

OPC is somewhat of an acronym embedded in an acronym. OPC stands for "OLE for process control"; in turn, OLE stands for "object linking and embedding." OPC is based on Microsoft's OLE/COM technology which essentially allows Windows programs to communicate with hardware devices.

OPC operates in a client–server approach. The OPC server converts the communications protocol of a hardware device into the OPC protocol. The OPC client software uses the OPC server to obtain data or send commands to the hardware device. The OPC client software, typically a human–machine interface (HMI), allows the client to communicate to the hardware device.

OPC is an open standard. Software vendors simply include OPC client capabilities in their products and they become compatible with hardware devices.

An OPC server has subsystems addressing functions such as data access, alarms and events, and historical data. The OPC server can interact with other applications such as Microsoft Excel, a web browser, or any ODBC database.

One analogy to an OPC configuration may be the use of printer drivers on PCs. Rather than have each application on a PC have a driver for a printer, one driver is used for all applications. This eliminates duplication, inconsistencies and conflicts.

Interoperable Smart Building System Databases

Each technology system in a smart building has some sort of data or database associated with its operation. Such data may also be required by another technology system, may be partially duplicated in another technology or management system, or may be needed by a business administration system. Major database standards allow for the access to or transfer of database information within smart building systems.

STRUCTURED QUERY LANGUAGE

The Structured Query Language is a standard (ISO and ANSI) that defines rules for the definition, structure, operation, manipulation, and management of relational databases. The first SQL standard was adopted in 1986 and there have been several additions, expansions and modifications of the standard since then. SQL is a vital and integral part of open network architecture necessary for smart buildings because it allows for databases from different manufacturers to interoperate and exchange data (Fig. 2.9).

IBM is recognized for developing the initial SQL format and rules that were later used to draft the SQL standard. Other major manufacturers have adopted the standard and market products compliant with SQL, including Microsoft, Sun, Oracle, and others. Many of these manufacturers have proprietary add-ons or extensions of the SQL standard. SQL can run on a variety of hardware (PCs, servers, mainframes), a variety of networks (local, wide, enterprise), and a variety of operating systems (MS Windows, UNIX, Linux, Mac).

SQL uses a row-and-column structure much like the spreadsheet applications used by many PC users. The initial purpose of SQL was to make it easy to query databases, but SQL has evolved to a full complement of programming, security, and management tools. Users can query data and programmers can program with simple sentences. Because the programming and user interfaces are simple and intuitive, data are more accessible and usable.

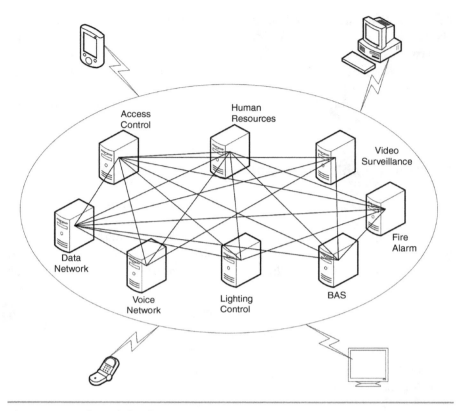

Figure 2.9 Shared databases.

OPEN DATABASE CONNECTIVITY

Other software applications typically work with SQL databases to provide additional functionality, connectivity, interoperability, and access. One of these is the Open Database Connectivity (ODBC) interface (Fig. 2.10).

ODBC allows ODBC-compliant software applications to access ODCB-compliant databases through a middle layer of software known as database drivers that resides between the application and the database. Developers of the application do not need to know the specific database the application will use. A software application can access several databases with multiple drivers.

ODBC is an SQL-based interface developed jointly by Microsoft and an industry working group called the SQL-Access Group but it is supported by other major manufacturers of SQL databases as well. This interoperability is important for

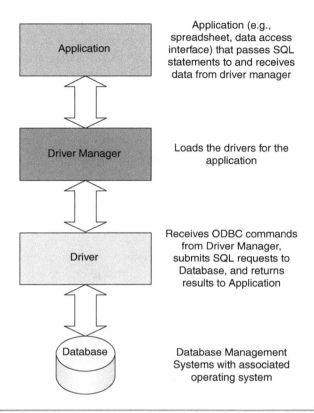

Figure 2.10 ODBC configuration.

open, integrated smart building systems. For example, data in a spreadsheet can easily be imported or exported to an ODBC-compliant database.

Power over Ethernet

When the first wave VoIP telephones and WAPs were installed they required two cables—one telecommunications cable to connect to the data network and another cable to a local power outlet. Older telephone systems never required local power for telephones because they were powered from a central telephone system, typically a private branch exchange (PBX). Many locations for VoIP telephones and wireless access points did not have local power available and power had to be installed at considerable cost. So the need for electric power for the first VoIP telephones and WAPs undercut the attractiveness and cost effectiveness of those technologies.

That shortcoming of especially the VoIP technology spurred the effort to supply power to the telephones over the twisted pairs in the telecommunications network cable. The result was that in 2003 an IEEE standard was published allowing low voltage power, 48 VDC, to be transmitted over Category 3 and Category 5 twisted-pair cable. With the current IEEE 802.af standard, the maximum that can be delivered to a powered Ethernet device is 15.4 watts. After counting losses, about 13 watts is the nominal power delivery available. Additional losses will occur with the use of switch mode power supplies.

The power is transmitted over the pairs of the Category 5 cable in one of two ways. For 1000BASE-T the power is delivered over the signal pairs. This method of powering is called "phantom power" (Fig. 2.11). For 10BASE-T and 100BASE-TX Ethernet the power can be transmitted over the two idle pairs. Transmitting the power over an idle pair is called "galvanic injection" (Fig. 2.12).

The device supplying power is called power-sourcing equipment (PSE), and the device being powered, such as the VoIP telephone or WAP, is called the powered device (PD). The PSE determines the method (galvanic injection or phantom power) used to power the PD. The PSE can be a POE-enabled network switch (referred to as an "endspan") or a device that injects power between a network switch not POE enabled and the PD (referred to as a "midspan"). Midspans allow users without POE-enabled network switches to make their networks POE enabled without procuring new network switches.

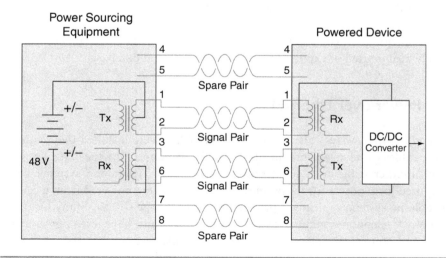

Figure 2.11 POE using the signal cable pairs.

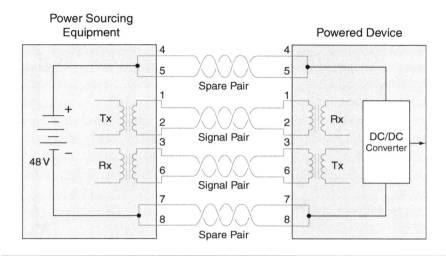

Figure 2.12 POE using the spare cable pairs.

The current standard is sufficient to power many low-powered devices. However, the power rating is insufficient to power devices such as video surveillance cameras using pan, tilt, and zoom (PTZ) capabilities and video telephones. A second standard, IEEE 802.at, also known as "POE Plus," is now being developed and will offer higher power levels although still within the low-voltage range.

Expectations are that the standards for POE Plus will have a power rating of at least 30 watts, basically doubling the existing standard. One concern with the development of the standard is that higher power generates more heat on the cable pair(s), resulting in an increase in the attenuation of the cable; the result could be that cabling runs be shortened to comply with the standards of ANSI/TIA/EIA-568.

POE products started with VoIP telephones and wireless access points but have rapidly expanded into hundreds of certified products. These include surveillance cameras, clocks, intercom systems, paging systems, access control card readers, time and attendance systems, and touch-screen flat panels. Descriptions of some of POE's numerous benefits follow.

Lower cost—The cost of a typical power outlet includes conduit, wire, a backbox for the outlet and the labor of an electrician. If a power outlet does not have to be installed for the device the cost of installation and construction is reduced. Some organizations have saved $350 to $1000 per location by not having to install power. Others have estimated that the average cost to provide power to a device is about $864 while the cost of a POE network port is $47 to $175.

Another potential cost savings is related to energy consumption. Several network equipment manufacturers have developed software that can turn POE devices on or off, or reduce the power to the devices, thus reducing energy consumption and costs.

Increased reliability—POE centralizes power distribution. Instead of a power outlet at each local device, power is now distributed from the telecommunication rooms somewhat like the older centralized telephone systems. Centralized power makes it easier to provide uninterruptible and emergency power for critical hardware thus increasing system reliability and uptime.

System management—POE allows the end device to be monitored and managed. Network switches provide management tools such as the Simple Network Management Protocol (SNMP) which allows staff to manage the end devices including the power to the end device.

Easier moves, additions, and changes—POE allows for slightly easier building renovations and rearranging of spaces since devices only need one cable. Installation of devices on walls or ceilings and temporary installation setups are easier.

International applications—POE is an international standard and deployed worldwide allowing manufacturers to avoid supplying different power cords for different countries.

Safer—When POE is used, lower-voltage distribution is used to power devices and less high voltage is used throughout a building. This results in a safer environment.

Chapter 3

Heating, Ventilating, and Air Conditioning Systems

Overview

Heating, ventilation, and air conditioning (HVAC) systems maintain the climate in a building. In other words, HVAC systems control the temperature, humidity, air flow, and the overall air quality. A typical system brings in outside air, mixes it with air returned from or exiting the system, filters the air, passes it through a heating or cooling coil to a required temperature, and distributes the air to the various sections of a building.

The HVAC system not only makes the building comfortable, healthful, and livable for its occupants, it manages a substantial portion of energy usage and related costs for the building. In maintaining the building's air quality, the HVAC system must respond to a variety of conditions inside and outside the building (including weather, time of day, different types of spaces within a building and building occupancy), while simultaneously optimizing its operations and related energy usage. The HVAC system is also critical in controlling smoke in the event of a fire.

HVAC systems in commercial and institutional buildings are very different from those used in typical residential housing. Larger buildings have a greater density of people, lighting and other equipment, all of which generates more heat. The result is that air conditioning, or the recirculation of air, becomes more important than providing heat, depending on the local climate. Although there may be a centralized HVAC system in commercial and institutional buildings, sections of large buildings have different HVAC needs or thermal loads depending on how the space is utilized.

An HVAC system having a single control thermostat serves one zone of a "thermal load." Most large buildings have multiple zone systems, with air supplied to each zone specifically addressing its needs and thermal load. For comparison on a smaller scale, a two-story home may have two zones, one for the lower floor and one for the upper floor, with one heating and cooling unit for each floor. The upper floor may have a higher thermal load and may require more cooling than the lower floor.

Components

HVAC systems can be very complex, consisting of many components. The major components include boilers, chillers, air-handling units (AHUs), air terminal units (ATUs), and variable air volume equipment (VAV).

BOILERS

Boilers are used to heat air (Fig. 3.1). However, because of the general increase in the efficiency of HVAC systems, many simply "recover" wasted heat produced from the chiller, another major component in an HVAC system, or use smaller-scale versions of traditional boilers to generate heat.

Boilers heat air in the following manner: a fuel (typically propane or natural gas) is combusted, and the resulting heat is used to heat water. The hot water or steam is piped through the building to radiator units where air is forced over them, moving heated air through the ducts and into the rooms.

Figure 3.1   Commercial boiler.

Whether steam or hot water is used as the heat transfer medium depends on the building's heating requirements. Heat transfer is simply the passage of thermal energy from a hot to a colder body. Hot water systems are usually more efficient and less susceptible to corrosion than steam systems. Steam systems are typically used in situations where large amounts of heat are required, such as centralized heating plants, but they require much more maintenance than hot water systems.

Boilers are available in two main categories: conventional units and condensing units. Condensing units allow the water vapor produced during the combustion of whatever hydrocarbon fuel is used to produce heat to condense. Condensing units typically have efficiencies of over 90% and are more energy efficient than conventional units. Enough heat can be extracted from condensing units that the exhaust gases are typically cool enough to be pumped through PVC piping.

Conventional boiler units are typically made of materials that cannot handle the corrosive properties of the condensing gases, and therefore that heat becomes waste. Conventional boilers can be retrofitted with a stack gas economizer, a device that captures some of the exhaust heat from the combustion gases and transfers it to the incoming water to the boiler, thereby raising the boiler's efficiency.

Boilers are also categorized in terms of heating methods: fire tube and water tube. Fire tube boilers transfer heat from combustion gases to the water using a series of straight tubes surrounded by water. The hot gases flow through the tubes and transfer heat to the surrounding water. Whereas, water tube boilers are composed of tubes housing flowing water surrounded by combustion gases that transfer heat to the water in the tubes. Water-tube boilers are capable of achieving higher capacities than fire-tube boilers because water or steam pressure can be contained within the tubes. Combined heat and power (CHP) systems are boilers that provide electricity while also providing heat for a building, but can be costly to install.

CHILLERS

Chillers, or air conditioners, utilize heat exchanges and circulate fluid or gas to cool the air that is passed through the unit. Chillers are often located in a mechanical area at ground level, or in a central plant in a campus environment. Chillers cool air by removing heat using the refrigeration or vapor-compression cycle (also known as the reverse-Rankine cycle), which consists of compression, condensation, expansion, and evaporation (Fig. 3.2).

A refrigerant in vapor form is initially compressed in a compressor, reducing its volume and increasing its temperature. It is then pumped to a condensing unit, where the refrigerant is cooled and condensed into a liquid. This liquid is then pumped to the indoor evaporator unit, where it is passed through

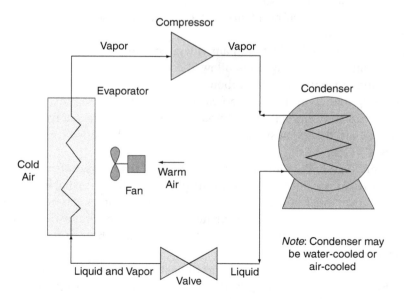

Figure 3.2 Vapor compression cycle.

evaporator coils that remove heat from the building. The hot air in the building is passed over the evaporator coils, adding heat to the refrigerant and removing heat from the air, which is recirculated back into the building. The added heat turns the refrigerant to vapor, which is sent back to the compressor, completing the cycle.

Chiller condensers remove heat from the system via cooling air, cooling water, and evaporation. Air-cooled systems are typically found in residential homes and commercial buildings where the cooling load is less than 100 tons (Fig. 3.3). The air-cooled condenser is comprised of coils that house the flowing refrigerant and maximize convective surface area (the area for the transfer of heat), and a forced air source (typically a fan) that uses convection to extract heat from the refrigerant and remove it from the system altogether.

Water-cooled systems are used for buildings that require large cooling loads, and typically have higher efficiencies than air-cooled systems. (Fig. 3.4). Instead of using air to remove the heat from the refrigerant, water is used to extract heat. Once the heat is extracted the water is then pumped to a cooling tower, where the heat is rejected back into the atmosphere and the water is then pumped back to the condenser (Fig. 3.5). Cooling towers reject heat by using an air stream to evaporate a portion of the incoming water, thereby cooling the rest of the incoming water. The heat transferred to the air causes it to rise, flowing out of the top of the tower and into the atmosphere.

Figure 3.3 Air cooled chiller.

Figure 3.4 Water cooled chiller.

Figure 3.5 Cooling tower.

Water-cooled chillers have higher efficiencies than air-cooled chillers because they reject heat at the wet-bulb temperature (which takes into account humidity and radiation), rather than the dry-bulb temperature at which air-cooled chillers reject heat. They are also smaller than air-cooled chillers for the same cooling output because its condenser requires less surface area and does not use fans, which also significantly reduces noise levels (Fig. 3.6).

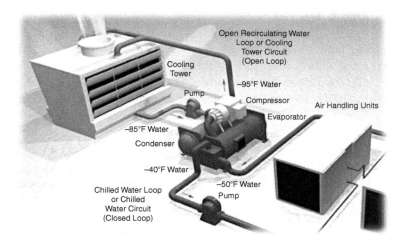

Open Recirculating Water
Loop or Cooling
Tower Circuit
(Open Loop)

Cooling
Tower

Pump

–95°F Water

Compressor

Air Handling Units

Evaporator

–85°F Water

Condenser

–40°F Water

Chilled Water Loop
or Chilled
Water Circuit
(Closed Loop)

–50°F Water
Pump

Figure 3.6   Water-cooled AC system.

Evaporative condensed chillers operate essentially as a smaller and more efficient version of a water-cooled system; they use the same evaporative cooling provided by a cooling tower. The cooling is achieved using a recirculating water system, which continuously wets the condenser tubes while fans blow air over them, evaporating the water and thereby moving (rejecting) the heat to the atmosphere.

Although this also uses air to move heat out of the condenser just as an air-cooled system would, an air-cooled condenser is less efficient because it draws ambient air over the condensing surface and rejects heat at the dry-bulb temperature. Evaporative condensed chillers use considerably less water than water-cooled chillers, thus reducing the operational costs.

For each type of chiller there are four compressor subcategories: reciprocating, centrifugal, screw driven, and absorption. Reciprocating, centrifugal, and screw-driven chillers are powered mechanically by electric motors, steam, or gas turbines while an absorption chiller is powered by heat and uses no moving parts. Reciprocating compressors use pistons driven by a crankshaft and are typically used for delivering small amounts of refrigerant at high pressure. Centrifugal compressors use centrifugal force to compress air and are used for delivering large volumes of refrigerant at low pressure. They are widely used in industry because they are energy efficient and have few moving parts. Screw-driven compressors use two opposing rotating screws to compress the air between them.

AIR-HANDLING UNITS

Air-handling units (AHUs) provide warm or cool air to different parts of a building, using chilled water to cool the air or steam or hot water to heat the air (Fig. 3.7). An air handler is usually a metal box containing a blower, heating and/or cooling elements, filter racks or chambers, sound attenuators, and dampers. Large AHUs for commercial use contain coils that use heated water provided by a central boiler and chilled water from a central chiller. Small AHUs usually contain a fuel-burning or electrical-resistance heater and an evaporative chiller that are integrated in the unit itself.

The AHU draws air in, passes the air over heating and cooling coils, and then forces it through air ducts. The AHUs have many of the networked points of the HVAC control system to manage air flow, heating, cooling and filtering. They can serve a building, a single floor on a building, or multiple floors of a building. If the AHU is serving multiple zones, each zone typically gets local control by having its own air premixed at the AHU. Some AHUs use no ductwork at all, recirculating the air in the space served.

Smaller AHUs, sometimes called blower coil or fan coil units (FCUs), can consist of only a coil, fan, and air filter, and operate using no outside air. Fan coil units are typically found in places where cooling requirements are small or on a room-by-room basis, such as hotel rooms and apartments. Larger AHUs, known as makeup air units (MAUs), operate using only outside air. One of the most common types of AHU is the roof-top unit (RTU), also known as a unitary air conditioner. The condensing unit for an RTU is on the roof of the building, with the cooling coils inside the AHU contained in the building. RTUs are most commonly used for one-story commercial buildings.

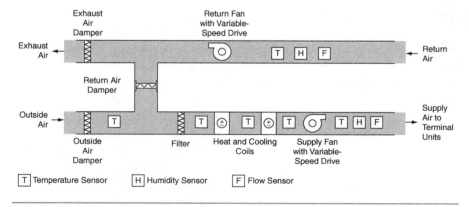

Figure 3.7 Typical air-handling unit.

AIR TERMINAL UNITS

Air terminal units (ATUs) address specific HVAC thermal loads or zones. Thermal loads in a space can consist of exterior loads (outside air temperatures increasing or decreasing) and interior loads (people, lighting, computers, and other sources). A thermal zone is a space or group of adjoining spaces in a building that have similar thermal loads.

Defining thermal zones in a building reduces the amount of HVAC subsystems needed because a single subsystem can usually handle an entire thermal zone. ATUs compensate for these thermal loads and zones by varying the air temperature, varying the air volume, or doing both. While constant air volume (CAV) systems provide air at a variable temperature and constant flow rate, variable air volume (VAV) systems provide air at a constant temperature and regulate the room temperature by changing the flow rate of the air into the room. VAVs can be pressure independent, where the flow is maintained constant regardless of the inlet pressure, or they can be pressure dependent, where the flow rate of the VAV is dependent on the inlet pressure and, typically, the position of its damper or speed of a fan.

VAVs are more energy efficient than CAVs. VAVs use less air volume resulting in less energy used for heating, cooling, and operation. Most residential HVAC systems are CAVs, while most new commercial buildings have VAV systems.

Efficiency

Air conditioner efficiency is essentially a ratio of the cooling capacity of the unit versus its required power input and is usually measured as follows: the energy efficiency ratio (EER), seasonal energy efficiency ratio (SEER), or coefficient of performance (COP). EER is a measure of an air conditioner's efficiency at its maximum air conditioning load, and is calculated by dividing the unit's cooling or heating capacity in BTUs per hour by watts of electrical input. EER is based on a constant 95°F outdoor temperature. A better and more widely used measure of efficiency is SEER.

Although it uses the same units as EER, SEER provides a more accurate measure of a unit's efficiency because it is taken as an average of various operating conditions and seasonal temperatures. It is designed to give the most accurate representation of actual operating conditions for the air conditioner. Because it is based on seasonal external temperatures SEER ratings depend on geographical location and thereby differ by area.

COP is a dimensionless ratio of the power output of a unit to the electrical power required to run it. EER can be converted to COP by dividing by 3.413 BTU/(W*hr).

Efficiency in HVAC systems has increased steadily over time due to advances in technology and rising energy prices. Efficiency has also increased due in large part to federal laws and national efficiency standards. The National Appliance Energy Conservation Act (NAECA) and the Energy Policy Act (EPAct) have established minimum-efficiency standards for furnaces, boilers, and packaged equipment to which manufacturers must comply, and the standards are updated every few years.

The American Society of Heating, Refrigerating, and Air-conditioning Engineers (ASHRAE) has created many of the standards that have eventually been adopted as requirements, code and law. The ASHRAE/IESNA Standard 90.1 provides the basis for most HVAC efficiency requirements in the United States. Compliance with the standard is a prerequisite for Leadership in Energy and Environmental Design (LEED) certification for new construction to demonstrate minimum energy performance of a building.

ASHRAE/IESNA Standard 90.1 includes efficiency requirements for equipment not included in federal laws such as EPAct and has been adopted as the commercial building energy code in many states. Because highest efficiency equipment can sometimes reach a cost-effectiveness limit there are federal and utility-sponsored programs that provide cash incentives for choosing such equipment.

STRATEGIES FOR MAXIMIZING HVAC EFFICIENCY

Heating, ventilation, and air conditioning account for a substantial portion of commercial building energy use and costs. In residential buildings, ventilation can be provided by windows and structural air leaks even if ventilation is not being provided by the HVAC system. Commercial buildings have mandatory ventilation requirements so the HVAC system provides most of the ventilation for the building at all times. Commercial buildings also have much higher thermal loads due to lighting systems, electronic equipment, and a higher density of people, and these loads grow larger as more floor space and levels are added to a building.

For all of these reasons increasing the efficiency of a building's HVAC system can result in dramatic savings over time. Numerous strategies have emerged to increase the efficiency of HVAC systems.

REDUCING LOADS

A major step to maximize HVAC efficiency in a building is reducing heating and cooling loads. Reducing heating loads is as simple as installing more efficient lighting and electronics. Incandescent lighting generates large amounts of heat while lighting an area. Modern energy-efficient fluorescent lighting provides the same and often better light quality with much lower heat output. Computer systems and servers also generate significant heat, and can be upgraded to more energy-efficient versions. Dark colors on the outside of a building increase the absorption of solar heat, increasing the thermal load of an HVAC system.

Reducing cooling loads can be accomplished by installing better insulation and more efficient windows and sealing air leaks. Interior spaces in a building often need cooling during times when the outdoor air temperature and humidity are sufficiently low to provide cooling without running refrigeration equipment. Economizers use controls and supply and return air dampers to provide "free" cooling by circulating outdoor air into a building when conditions allow.

EQUIPMENT SIZING

Correctly sizing HVAC equipment is extremely important. HVAC systems are often designed to handle the maximum heating and cooling loads possible for an area, keeping a building cool on the hottest days and warm on the coldest days. This leads to the HVAC system being oversized to handle just a few days out of the year. Oversizing HVAC equipment wastes energy and can also be uncomfortable for inhabitants by overheating or overcooling a space. Similarly, undersizing can lead to discomfort and underventilation, which can lead to high carbon levels and poor health for inhabitants.

Sizing is such an important step that it needs to be well thought out before a system is even designed. Buildings with floor areas less than 20,000 square feet typically use factory-built, air-cooled "unitary" (packaged) equipment; buildings from 20,000 to 100,000 square feet typically use multiple large packaged units; and buildings larger than 100,000 square feet and multibuilding campuses generally use water-cooled systems specifically built for that location.

Despite these "rules of thumb" the correct sizing of an HVAC system's cooling capacity should not be based purely on the size of a building. The envelope load (windows, walls, and roof), the internal thermal loads (lights, people, and equipment), and the ventilation load must also be considered. ASHRAE provides proven methods for calculating heating and cooling loads.

To improve efficiency an HVAC system must be considered as a whole system rather than a number of parts or components. For example, in a chilled-water system the chiller is the main component and the largest energy consumer. However, simply selecting a high-efficiency chiller does not guarantee a high performance system. Auxiliary equipment, such as fans and blowers, can also have substantial effects on efficiency. According to the U.S. Department of Energy, fans, and pumps account for 64% of the electricity consumed by a commercial HVAC system.

Inefficient auxiliary equipment and a poorly designed, sized and placed system can severely reduce overall system efficiency. For example, exhausts and intakes must have maximum airflow without restrictions, condensing units should be spaced far enough apart so as to prevent the recirculation of heat, and outdoor units should be located close to indoor units.

HVAC SEQUENCE OF OPERATION

HVAC systems operate based on a predetermined sequence of operation. The sequence of operation is critical to the efficiency and effectiveness of the HVAC system. It is the sequence of operation that will determine how system components interact. As an example, an HVAC system is typically triggered when the thermostat senses that a space's air temperature differs from the thermostat's set temperature. When that happens, contacts in the thermostat are closed and control voltage is supplied to the control board terminals, which causes the blower in the AHU to start.

The AHU control board then supplies control voltage to the condenser, causing a device known as a contactor to close its contacts. This supplies power to the compressor, which then raises the pressure and temperature of the refrigerant gas in the system, forcing it through the condenser coils. In the condenser the refrigerant is cooled and condenses from a gas to a liquid and is then pumped indoors to the evaporator through a metering device that regulates the flow to the evaporator coil, causing the refrigerant pressure to decrease. Air is then circulated over the evaporator coil by the fan in the AHU, transferring heat from the air to the refrigerant. This heat raises the temperature of the refrigerant above its saturation point, changing it back into a gas. The refrigerant gas is then pumped back to the condenser where the process repeats itself until the thermostat senses that the temperature is at the desired level, opening the contacts and shutting off the control voltage supplied to the furnace or the air conditioner.

MAINTENANCE

The placement of certain components also affects how often a system will undergo maintenance. Maintenance of an HVAC system is critical to maintain its efficiency and performance and adequate space to perform such maintenance is crucial. When equipment is crammed into a space that makes it difficult for a technician to service, it is less likely that the equipment will receive regular maintenance. Routine maintenance such as replacing filters, and less routine maintenance, such as replacing compressors, fans, belts, shafts, bearings, and coils is essential to maximize a system's life-cycle efficiency.

DISPLACEMENT VENTILATION

The concept of "displacement ventilation" can drastically improve HVAC system efficiency and ventilation quality. Traditional ventilation mixes a turbulent stream of fresh air with waste air that has been exhaled by occupants of a space, creating a constant mix of medium-quality air. Displacement ventilation uses a slow-moving stream of fresh air from the floor to displace the waste air, which is forced to the ceiling and then out of the room through exhaust panels.

This creates two levels of air in a room, with cool fresh air in the occupied lower part of the room, and warm waste air in the unoccupied upper part of the room. This also creates natural convection, because as the cool air rises it cools the occupants and then the heat taken from them is expelled from the room. Displacement ventilation can eliminate the need for large HVAC equipment, not only reducing energy costs but initial and maintenance costs as well.

HVAC Controls

HVAC systems must control variable conditions of the system and its components. These conditions include liquid and gas pressure, temperature, humidity, the flow rate of liquids and gases and the speed and on/off state of mechanical equipment.

A number of instruments and terminal devices available in the field are used to gather data on the system and assist in controlling it. System controllers use input and data from sensor devices to make decisions about the system, and then, based on the input information, control actuator devices.

Sensors and transmitters include thermostats, liquid differential-pressure transmitters for pumps and chillers, differential pressure sensors for fluids and airflow, static pressure sensors, air-pressure sensors, and humidity sensors. An example of an actuator or operator is an actuator for a damper that is mounted to the damper shaft and triggers the start of the damper operation. That operation could be a temperature sensor detecting a high temperature and sending a signal to the controller, which results in the controller sending a signal to an actuator to engage a motor that opens or closes a damper or vent.

These devices may communicate to each other or to the controller with analog or digital signals. Analog inputs to a controller can be a continuously changing signal from an external device or sensor, such as a temperature sensor. Digital inputs to a controller are simply a two-state, on-off signal from external devices or sensors, such as a switch.

In much the same way, analog outputs from a controller are "proportional variable" signals sent by the controller to adjust an actuator or external control device, such as a valve actuator. Digital output from a controller is a two-state or two-position signal from the controller to an actuator, such as a control fan relay start-stop switch.

Most field devices and equipment for building automation systems communicate at low network speeds, typically transmitting at rates of less than 1 Mbps. The communications network for a building automation system is typically in a physical star or bus topology from the controller.

Older HVAC system controls were provided through electric power or pneumatic means. However, direct digital control (DDC) is commonly used in more complex HVAC systems. DDC allows for a system controller to compute the sequence of operations based on the digital input from system sensors. Although DDCs are digital controls, they are able to handle analog-to-digital and digital-to-analog conversions. Unlike electric or pneumatic controls, DDC can be programmed for any sequence of operations.

Controllers can, confusingly, be referred to in a variety of ways: master, slave, terminal, floor, and others. HVAC system network architecture typically consists of the following network levels:

- Management level
- System-level or building-level controllers
- Field-level controllers

MANAGEMENT LEVEL

The top level of an HVAC control system is the management level consisting of personal computers or multiple PCs connected via an Ethernet network. These operator workstations can communicate with, interrogate and control any of the controllers and devices on the network. The management level provides many functions:

- Administration and control of the HVAC system
- Programming for the system and other controllers, including operation sequences
- Display of system information

- System reports
- System scheduling
- Archive and analysis of historical data
- Backup of controller databases
- Alarm reporting and analysis
- Trend analysis

The HVAC system is usually managed by a server and operator workstation using standard operating systems, specific HVAC software applications, GUI interfaces, and web access. The HVAC control system may be interfaced or integrated with fire alarm, video surveillance, access control, and lighting control systems. The HVAC system is also a significant part of a facility management and maintenance management system, primarily for tracking, managing, and optimizing energy use.

SYSTEM-LEVEL OR BUILDING-LEVEL CONTROLLERS

The system-level or building-level controllers are networked back to the management level. In a campus environment, the building-level controllers are networked via a campus network to the management level of the HVAC control system. These controllers can manage HVAC equipment directly (typically, major components such as air-handling units) or indirectly through networked downstream, lower-level controllers. System level controllers handle the operations of all downstream field level controllers, collect and maintain data, and can operate as standalone units if communication is lost to the management level. System controllers have a peer-to-peer relationship with other controllers.

FIELD-LEVEL CONTROLLERS

Field-level controllers serve building floors, and specific areas, applications, and devices. Field level controllers are limited controllers in terms of both functionality and connectivity. Included in this group are DDC controllers, mechanical controllers, and application-specific controllers.

DDC controllers may support multiple applications, specific device networks, or a particular equipment component, such as an air handling unit. The DDC controller usually has onboard memory, an operating system, and a database. Both DDC and mechanical controllers perform control through control algorithms. For example, the controller may measure temperature or humidity in a specific area and based on the measurement, direct cooling, heating, humidifying, or dehumidifying to that area.

Some mechanical equipment, such as air-handling units or chillers, may be procured with a field-level controller and devices as part of the equipment. System controllers allow field controllers to communicate with other field controllers or a group of field controllers, and to access databases and programs. DDC controllers can also use remote application-specific controllers (ASCs) for devices such as VAV terminal units.

HVAC control systems are evolving to the utilization of smart building infrastructure. This is true of the HVAC system management level, and is evolving or trickling down to the network hierarchy at the system and field level. The adoption of ANSI/TIA/EIA-862 addressing–structured cable infrastructure for building automation systems allows for standard unshielded twisted pair copper cabling and fiber optic cabling to be utilized throughout an HVAC control system.

Although the IP network protocol may be used at the management level, BACnet or LonTalk are likely to be used at other levels. However, these other protocols recognize the dominance of the IP protocol and are either providing routers for transitioning from their native protocol to IP (LonTalk) or are migrating to IP with standards such as BACnet/IP.

Chapter 4

Lighting Control Systems

Overview

Facility lighting is needed to provide visibility for building occupants, aesthetic atmosphere for spaces or rooms and for life safety. It is estimated that lighting accounts for 30 to 40% of electricity usage and costs in a typical building. Therefore, unneeded and uncontrolled lighting in a building not only wastes energy but also increases facility operational costs.

In addition, lighting can affect other building technology systems such as the need for and costs of cooling spaces where lighting has increased space temperature. Lighting control systems provide lighting for occupants of the building as needed in an efficient manner, consistent with any applicable building and energy codes.

The need for lighting in a building varies by the type of building, spaces within the building, time of day, and occupancy of the building. Consequently, the control strategies and functions of a lighting control system reflect these variables and primarily involve the following:

- **Scheduling**—A control system may have a predetermined schedule when lights are turned on and turned off.
- **Occupancy sensors**—For spaces in a building where occupancy is difficult to predict (such as meeting rooms or restrooms), lights may be turned on and off based on a lighting control system device sensing occupancy.
- **Daylight**—To reduce the need and cost of lighting spaces a control system utilizes natural light as much as possible. This is sometimes called "daylight harvesting" or "daylighting."
- **Window coatings**—"Spectrally selective" window coatings, designed for hot climates with large amounts of solar radiation, work by selectively filtering out frequencies of light that produce heat while minimizing the loss of visible light transmission.

The lighting control system distributes power to the available lighting units in a typical fashion, but inserts digital control and intelligence in many, if not all, of the devices controlling the lighting such as the circuit breaker panel, wall switches, photo cells, occupancy sensors, backup power and lighting fixtures. The control system significantly increases the functionality and flexibility of the lighting system by providing digital control and intelligence to the end devices. For example, a reconfiguration of lighting zones is accomplished through software rather than the physical recabling of the lighting zones. In additional, intelligent end devices allow more focused application of lighting control needs and strategies to specific spaces in the building.

System Control

The heart of the lighting control center is typically a server that is web-enabled and interconnected to other facility technology systems, a workstation with a GUI interface and client software for system administration. The networked system allows any authorized individual, including tenants or other occupants, to adjust their lighting through the network or a web browser.

One approach to the lighting control system is the use of intelligent controllers. These controllers are distributed throughout a facility and manage downstream relay panels. The controllers and the system server are networked via an Ethernet network, usually sharing schedules and overrides. The controller may have a user interface panel which can be used instead of a system workstation to program and monitor the lighting control system (Fig. 4.1).

System controllers may be modular to allow for growth. The controller may also have several communications interfaces such as an Ethernet port and ports for RS-232 and RS-485 communications. The system controller communicates with each of the panels through an Ethernet connection or a BACnet, LonTalk or Modbus protocol that is routed to a backbone IP network.

Another emerging networking approach for lighting control systems is the distribution of the intelligence and controls further downstream to each device. Typically this would be a network interface for each lighting ballast. This approach centralizes the control to the network server and allows for network interfaces to specific devices.

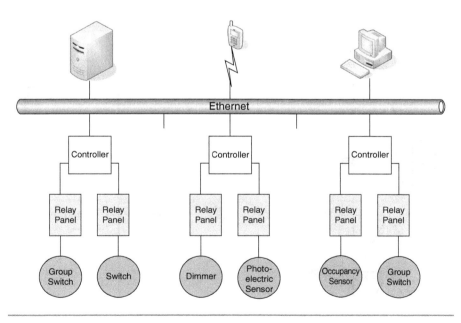

Figure 4.1 Typical lighting control system.

RELAY PANELS

Relay panels are typically mounted next to the electrical circuit breaker panels. The circuit breaker panel feeds into the relay panel with the relays in the relay panel acting as a switching device for the circuit. Many relay panels can be fed by both 120 V and 277 V circuit-breaker panels and relay groups can be fed by different voltages in the same panel. Each relay can be individually programmed through the system controller or the relay panel (Fig. 4.2).

Relay panels provide line voltage control of the lighting loads; they allow for a single circuit to feed into several relays and for multiple circuit breaker panels to feed into a single relay panel. While relay panels can be programmed or controlled by a system controller, they can also operate without the system controller. The relay panels typically have status indicators for the relay outputs, dry contact inputs for program override purposes, and inputs for monitoring devices, such as photo cells and occupancy sensors.

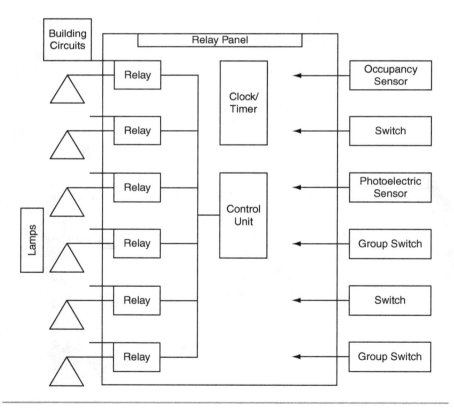

Figure 4.2 Lighting control relay panel.

In a multistory facility, there may be a relay panel on each floor controlling all the lights on the floor. Each room on the floor has a local switch and there is also a master switch for the entire floor. The master switch for the floor may be programmed to turn lights on at 7 AM and off at 6 PM; between 6 PM and 7 AM, the system may repeatedly perform an "off" sweep to turn lights out where the programming has been overridden.

OCCUPANCY SENSORS

Occupancy or motion sensors are devices that sense the presence or absence of people within their monitoring range. Unlike scheduling controls, occupancy sensors do not operate on a time schedule; they merely detect whether a space is occupied. They may be used in restrooms, utility rooms, conference rooms, coffee rooms, locker rooms and many other spaces. Typically, the sensor and a control unit can be enclosed in one unit, such as a wallbox, but for larger facilities, the sensor is tied to a relay panel.

The control unit or the relay is programmed to turn lights "on" when the presence of people is sensed by the motion detector, and may be programmed to turn the lights "off" if the space is unoccupied for a predetermined time period. The sensitivity of the sensor may also be adjustable (Fig. 4.3).

There are several types of motion sensors available, including passive infrared (PIR), active ultrasound, and hybrid technologies, such as combinations of PIR and active ultrasound, or PIR and audible sound. These sensors are typically used in locations such as hallways, lobbies, private offices, conference rooms, restrooms, and storage areas.

Ultrasonic sensors emit high-frequency sound waves and sense the frequency of the reflected waves as they return to the device. Movement in the area where the waves are emitted changes the frequency of the reflected waves causing the sensor to turn the lights on. Ultrasonic motion detectors provide continuous area coverage and are best suited for use in open areas such as offices, classrooms, and large conference rooms. Mechanical devices that produce vibrations or changes in airflow, such as HVAC systems, can however trigger ultrasonic occupancy sensors and cause lights to turn on.

PIR sensors detect radiation, that is, the heat energy that is released by bodies. They are labeled "passive" because they only accept infrared radiation and do not emit anything. PIRs operate in a line of sight and have to "see" an area, so they cannot be obstructed by open-area partitions or tall furniture. PIRs use a lens to focus heat energy so that it may be detected. However, the lens views the covered area through multiple beams or cones and may create coverage gaps. Any objects that prevent the sensors from "seeing" portions of its

Figure 4.3 Occupancy sensor.

designated area will cause the sensor to assume that the area is unoccupied and turn the lights off, even when it isn't.

Careful placement of occupancy sensors is required to prevent false alarms. Occupancy sensors should not be mounted in the direction of a window. Although the wavelength of infrared radiation to which the chips are sensitive does not penetrate glass very well, a strong infrared source such as a vehicle headlight or sunlight reflecting from a vehicle window can overload the chip with enough infrared energy to fool the electronics and cause a false alarm. A person moving on the other side of the glass however would not be "seen" by the device.

Devices should not be placed in such a position that an HVAC vent would blow hot or cold air onto the device surface. Although air has very low emissivity (emits very small amounts of infrared energy), the air blowing on the plastic window cover could change the plastic's temperature enough to "fool" the electronics.

Other technologies and approaches to motion detection include sensing audible noise. Hybrid sensors (PIR and ultrasound, PIR and audible) offer the most effective occupancy detection and have maximum sensitivity without triggering false detections.

The placement of occupancy sensors is key to their proper operation. Sensors can be mounted on walls or ceilings, and the use of multiple sensors can sometimes provide more accurate detection especially in large or irregularly shaped areas. Occupancy sensors must be able to detect motion in their assigned space while ignoring mechanical vibrations and other false signals. Any malfunctioning of an occupancy sensor can be dangerous, especially if the area is a stairwell or other location where illumination is important for safety. Occupancy controls are best used in applications where occupancy does not follow a set schedule and is not predictable.

DIMMERS

Dimmer modules manage low-voltage switch and line voltage output controls of the dimmer's lighting loads. Stand-alone dimmers typically have status indicators, analog inputs for photo-cell or occupancy sensors, diagnostics, and are able to optimize responses for various types of lighting fixtures. Dimmers can be used for specific spaces such as areas with audio visual presentations or throughout the total system for managing large facilities (Fig. 4.4).

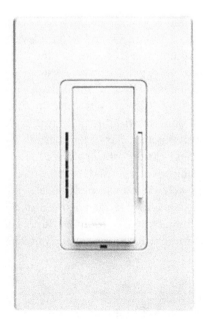

Figure 4.4 Dimmer.

Like occupancy sensors, dimmer switches are connected to a relay panel. Preset dimming controls from a relay panel provide predetermined dimming for several channels or loads. Presets are tamper-proof, that is, they will not allow anyone except authorized lighting control personnel to override the presets.

Dimming can be use to implement several energy-savings strategies. For example, lights can be dimmed when the demand for electricity exceeds a predetermined level, possibly as part of an overall load-shedding policy. Such reductions are typically unnoticeable by most users.

Another example involves fluorescent lamps. The light output of fluorescent lamps decreases over the life of the lamp (the expected depreciated output may be used as an initial design factor). Dimming can be used with new fluorescent lamps to produce the desired light level, then gradually manage the lighting level over the life of the lamps to produce both a constant level of light output as well as longer lamp life.

DAYLIGHT HARVESTING

Photoelectric controls are designed to strategically use daylight to reduce the need for artificial lighting, a process called "daylight harvesting." They may be located in perimeter offices, atriums, hallways or in areas with skylights. Ambient light sensors measure natural and ambient light then based on the amount of natural light, adjust the lighting to maintain a constant light level. In some spaces manual or automatic blinds, or other means of reducing the direct solar exposure glare, excessive light levels, and heat gain, can be used to supplement photoelectric controls. These may include motorized window shades or blackout shutters.

Proper daylight-harvesting design not only includes providing adequate daylight to an area but does so without undesirable side effects such as heat gain and glare. Successful daylight-harvesting designs will incorporate shading devices to reduce glare and excess contrast.

Window size and spacing, glass type, and the reflectance of interior finishes must be taken into account as well. Despite all of these design considerations, daylight harvesting provides little benefit without an integrated electric lighting system due to the increased thermal loads from the sun. The electric lighting and thermal loads must be reduced while simultaneously increasing daylight to an area.

BALLASTS

An electrical ballast is a device that limits the amount of current in an electric circuit. In electrical gas-discharge lights such as fluorescent and neon lights ballasts control the current flowing through the light.

Incandescent light bulbs produce light by running electricity through a metal filament inside a bulb which heats it and causes it to glow and emit visible light. When fluorescent lights are turned on electricity flows to two electrodes on opposite ends of the lamp causing them to heat up. The electrodes, which are very similar to a filament in an incandescent bulb, then become hot and emit electrons that collide with and ionize noble gas atoms inside the bulb. This creates a voltage difference between the two electrodes causing electricity to flow between the two electrodes through the gas in the tube. These gas atoms become hot vaporizing the liquid mercury inside the tube. The mercury vapor then becomes excited and emits ultraviolet light which hits a white phosphor coating that converts the ultraviolet light into visible light.

Due to an effect known as "avalanche ionization" (in which the gas continues becoming more excited and higher intensity light would be emitted until the light failed) a device is necessary to regulate the electricity flowing through the bulb. Thus, it is necessary to have a ballast to regulate current through the gas. Modern ballasts supply the electricity needed to start the lamp and produce light, then regulate the current so that the lamp will produce the desired light intensity (Fig. 4.5).

There are two main types of ballasts: magnetic and electronic. Magnetic ballasts use electromagnetic induction to create the voltages used to start and operate fluorescent lights. They contain copper coils that produce electromagnetic fields to control voltage. Magnetic ballasts, which have been used in fluorescent lights since their origin, are considered outdated and are being phased out by newer electronic ballasts. Electronic ballasts use solid-state circuitry, rather than magnetic coils, to control voltage to the lamp, making them more energy efficient.

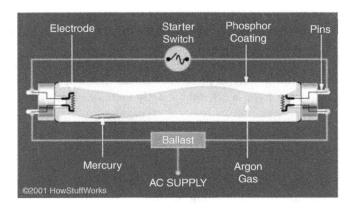

Figure 4.5   Ballast.

One of the main problems with magnetic ballasts is that while they can control the current to the light, they cannot alter the frequency of the input power. Because of this the lamp illuminates on each half-cycle of the input power causing the lamp to flicker and produce a low humming noise. This flicker can cause eye strain and headaches in some people and the humming can be bothersome and distracting. Electronic ballasts control input frequency, thus eliminating these problems. Another advantage of electronic ballasts over magnetic is that one electronic ballast can control more than one lamp, allowing for multilamp fixtures to be controlled by a single ballast.

Fluorescent ballasts come in three different types: rapid start, programmed start, and instant start. Rapid-start ballasts start lamps by simultaneously providing voltage to the electrodes and across the lamp itself. As the electrodes become hotter and emit more electrons, less voltage is required for the lamp and eventually the cathodes will become hot enough to ignite the lamp on their own.

Programmed-start ballasts are more advanced versions of rapid-start ballasts. They have preprogrammed start-up sequences designed to give superior longevity to lamps. Rather than supplying simultaneous voltage, programmed-start ballasts first apply voltage to the electrodes to heat them up for a short interval, then apply voltage to the lamps. This helps avoid a common problem in fluorescent lights called "tube blackening" which occurs when the electrodes are damaged from voltages without sufficient heating. Programmed-start ballasts have the longest lamp lives and are best used in locations with lights that are constantly being turned on and off.

Instant-start ballasts start lamps by providing high voltage directly to the lamps without preheating the electrodes at all. Because there is no heating time, light is produced within 50 milliseconds, thus giving it the name "instant start." Instant-start ballasts have the highest energy efficiency of any ballast and the lowest cost however they suffer from emissive material defects like the rapid-start ballasts. Instant-start ballasts are best used in lights that are not turned on and off very often.

For certain applications where a light is intended to be constantly turned on and off, such as a flashing light, there are ballasts that will keep the electrodes heated even when the light is off, thus greatly increases lamp life span.

INTEGRATION INTO BUILDING AUTOMATION SYSTEMS

Lighting systems provide a life safety function, assisting in security or lighting evacuation pathways from a building. Lighting systems may be integrated with fire alarm systems, security systems or emergency power generators. In the case of a fire alarm or loss of normal power the lighting control system may turn on key emergency lighting fixtures.

Data and information from a lighting system are also an integral part of an overall energy strategy and at a given facility or business level, need to be considered with HVAC systems, metering and building plug loads. Monitoring the number of hours that the lights are operated and the number of times that lights are turned on provides information to estimate lamp life, which can be used to schedule group relamping.

Lighting control systems may use several different communications protocols. One communication protocol called the Digital Addressable Lighting Interface (DALI) was specifically developed for ballasts and relay switches in the 1990s. For a DALI implementation each lamp uses a dimming ballast and each lamp has its own network address. DALI uses a two-wire, low-voltage wiring scheme in a bus topology with the bus providing both power and control signals to the ballast.

DALI is an open-source protocol based on standard IEC60929 which specifies performance requirements for electronic ballasts. Each DALI controller (called a busmaster) can handle up to 64 addresses and 16 groupings. Because the DALI protocol is solely for use by lighting control systems, the integration of a DALI lighting control system to other building automation systems requires a protocol translation with systems using BACnet or LonWorks. DALI has been extended into shading control and wired network connectivity has been supplemented with wireless connectivity.

Overall lighting control systems are increasingly using structured cable, Ethernet, and TCP/IP protocols at least at the higher levels of the control system—all of which are foundations for a smart building.

Chapter 5

Electric Power Management Systems

Overview

A facility's electric power management system (EPMS) monitors the power distribution system for usage and quality. The EPMS, together with the HVAC system and the lighting control system, are integral to overall energy management aimed at controlling usage and costs. In addition, the EPMS is a tool in managing and ensuring the quality of the power, that is, a power source free from surges, sags, and outages that may affect the facility's reliability and safety.

The EPMS monitors the electrical distribution system, typically providing data on overall and specific power consumption, power quality, and event alarms. Based on that data the system can assist in defining, and even initiating, schemes to reduce power consumption and power costs. The schemes "shed" power and are triggered by predetermined thresholds such as certain levels of power demand or a particular time of day when resource consumption is high. The EPMS can manage event alarms, calculate usage trends, track and schedule maintenance, troubleshoot, and "bill back" metered power usage to specific users or tenants.

Typical systems monitor the power service entrance of a building or campus, switchgear, generators, network protectors, switchboards, panel boards, uninterruptible power supplies (UPS), emergency power generation, and more. The components of an EPMS include monitoring and control devices (Fig. 5.1).

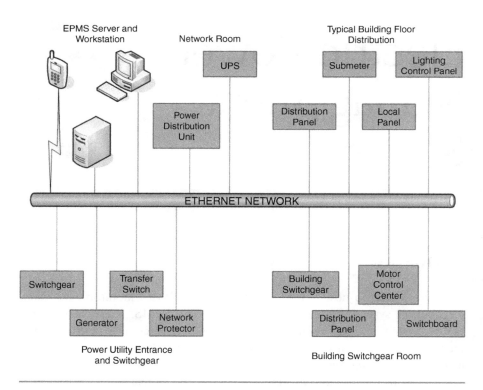

Figure 5.1   Electric power management system.

Monitoring

The EPMS monitors the electric loads of major equipment. Inputs to the monitoring unit can be current or potential transformers as well as other sensors or monitoring devices. Current transformers are used to provide information regarding electric current while potential transformers are used to provide information for electric voltage meters.

Typically the monitoring units are microprocessor based, have onboard memory, and can be programmed or have preset factors for monitoring, testing, and reporting sets. Some monitoring units can display locally and/or communicate via the EPMS network.

The monitoring unit may have inputs for specific distribution components such as electrical breakers. Addressable relays may be used for sensing and the monitoring of devices where information is simply in an open/closed or on/off state. The monitoring of critical equipment such as switchboards and switchgear may include digital metering and the capability to monitor power quality. For power quality most EPMS implementations adhere to the IEEE categories for power quality problems, including sags, swells, harmonics, interruptions, under- and over-voltages and transients.

Display Units

Display units typically connect to the monitor of an electrical load or equipment. The displays can be local, specific to particular equipment, or allow for viewing and monitoring of multiple loads and devices. Some larger display units may connect to multiple monitors and are able to communicate with an operator's workstation via a standard data network.

Central Operator Workstation

A central operator workstation for the EPMS is usually a personal computer with special application software. It uses data from the system components to analyze and take action regarding power usage in a facility. The operator workstation can have the following features:

- Distribution system graphics
- Real-time reports
- Trend reports

- Historical reports
- Alarm reporting
- Analysis of electric power waveforms
- Determination and initiation of power load-shedding strategies
- Communication with HVAC and lighting control systems
- Usage billing software

EPMS implementations have evolved to the infrastructure foundations of smart buildings. Backbone EPMS network connections can use standardized, unshielded, twisted-pair copper and fiber optic cables, supplemented with shielded, twisted-pair copper cable. Physical topologies can be bus, star or daisy-chain configurations.

Many of the lower-level network communications use RS-485 communications while higher network levels use TCP/IP protocols. EPMS network components and devices connect to the Ethernet network directly or through connectivity to a network router or gateway, essentially converting or encapsulating other network protocols such as Modbus. Databases for EPMS environments typically adhere to SQL standards. Systems also allow for connectivity to remote-client personal computers and PDAs. The EPMS is an important facility operational tool and a critical element in a smart building.

Demand Response

An EPMS is particularly useful in allowing for demand response. Demand response refers to mechanisms that manage the consumption of electricity based on electricity supply availability and pricing. An EPMS can initiate power consumption reduction schemes when certain levels of power demand are reached.

At certain times of the day known as the "peak" hours electricity demand spikes and power plants fire up their peaking power units. This peak electricity is generally more expensive than the base electricity that is produced because it is usually provided by a quickly fired source such as a gas turbine. This is because it is not cost effective to size a power plant based on the peak load; the unnecessary capital costs far outweigh the more expensive peaking fuel. Thus, it is in the electricity provider's best interest to reduce the amount of peak electricity they have to produce.

Demand response can be explained using a quantity (Q) versus price (P) graph (Fig. 5.2). Price elasticity of demand (PED) is a measure of how consumers react to a change in the price of a certain commodity, in this case electricity.

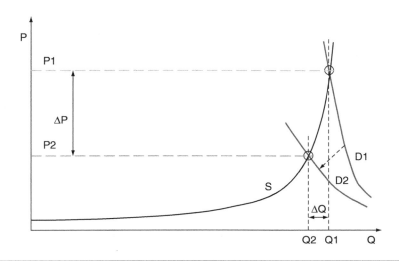

Figure 5.2   Demand response.

Under inelastic demand conditions (D1) consumers will pay nearly any price for electricity; extremely high prices (P1) may result in a strained electricity market. If demand response measures are employed the demand becomes more elastic (D2), meaning that consumers are more in control of electricity prices and will not pay more than they want to. In this case, a much lower price (P2) will result in the market.

Whether the consumer is purchasing electricity in a regulated or deregulated market is a major factor in determining how demand response is considered. In some government-regulated markets consumers have no financial incentive to reduce their peak load since their rate will not change. In this type of market, utilities often provide programs and incentives for large commercial customers to reduce peak demand. In a deregulated market, consumers are purchasing electricity at a market price that varies throughout the day so they are forced to lower or defer their peak demand to reduce costs.

The electrical demand of a building can be reduced by the EPMS in response to a request by a utility or to market-price conditions utilizing a method called load shedding. Load shedding is the reduction of electricity usage from building systems such as lighting and HVAC systems according to a preplanned load-prioritization scheme. Whenever peak hours have passed, electricity usage returns to normal.

An alternative to load shedding is on site generation of electricity during peak hours to provide the electricity needed past the base load.

Special equipment is needed for load shedding. In addition, the individual energy-consuming components must also be compatible for load shedding. For example, the electrical ballasts for the light fixtures in a building must contain internal devices that allow them to receive signals from the EPMS or have the EPMS trigger the lighting control system.

Demand response can significantly reduce peak electricity rates and volatility. A study carried out in 2007 by the Brattle Group showed that even a 5% drop in peak demand in the United States could produce enough savings in generation, transmission, and distribution costs to eliminate the need for installing and running some 625 peak-power plants, creating $35 billion in savings over the next two decades.

Electricity Usage Metering and Submetering

The use of "smart meters" is becoming more prevalent as electricity prices rise. A smart meter identifies consumption in more detail than a conventional meter, sends data back to the local utility for monitoring and billing purposes, and may provide information and data to the consumer directly. This makes possible real-time pricing for all types of users, as opposed to fixed-rate pricing throughout the demand period.

An electric meter is a device that measures the amount of electricity consumed at a particular location. In order for a utility to accurately bill a customer a meter is needed to record consumption. When billing a customer, utilities record the electricity usage as well as the time when the electricity was used, since prices may vary throughout the day (Fig. 5.3).

Meters can be incredibly useful to a utility customer. Meters provide information allowing customers to control electricity consumption and demand, determine typical consumption, evaluate opportunities for improving energy efficiency and confirm savings from building and/or building systems improvements. Conventional meters cumulatively measure, record, and store aggregated kWh data which are used for billing or energy management purposes. Newer smart meters allow utility customers to monitor demand, power outages and power quality in real time.

Smart meters have many advantages over conventional meters. Smart meters have the capability to measure and record interval data and communicate the data to a remote location. Smart meters allow utilities to introduce different prices for electricity based on the time of day and the season, which can be used to reduce peaks in demand. Smart meters can locate power quality issues such as transients, voltage disturbances, power factors and harmonics.

Figure 5.3   Electric meter.

Transients can cause premature failure in electronic equipment, improper power factors can result in surcharges from utility companies and high harmonics can shorten the life of transformers. Once the smart meter locates these issues they can be dealt with accordingly.

Meters are most commonly used for billing; however, the data provided by a conventional meter is usually not very detailed, often consisting of nothing more than a monthly total of energy usage. Many utilities are moving toward time-of-use rates, that is, they charge more for energy use during times of peak demand. They will also often provide an incentive for utility customers to shift demand to off-peak periods. Because conventional meters only provide aggregate readings, time-of-use rates require smart meters. EPAct 2005, Section 1252 requires that utilities in a state that mandates time-of-use rates must provide smart meters to their customers.

Real-time pricing, also known as dynamic pricing, makes even better use of smart meters. Time-of-use rates are predetermined by the utility and provided to customers, allowing the customer to adjust or shift their load accordingly. Real-time rates change by the hour, based on the utility's actual cost of generating and/or purchasing electricity at that particular time. Being able to adjust energy usage in response to these price changes can save electricity customers a great deal of money. Smart meters allow customers to monitor these price changes and change their consumption if they choose.

Submetering is extremely helpful for understanding the way that energy is used in a particular building or area. Utilities often install a single electricity meter for an entire group of buildings which serves their purposes but does not provide any information to the customer about load distribution among different areas or buildings.

Without submetering a utility bill is often allocated according to the square feet of occupied building space. With the use of submeters, payments for utilities are based on measured usage, so there is an incentive for building occupants to conserve utility resources.

A submetering system typically includes a "master meter" and a number of submeters. The master meter is owned by the utility supplying the electricity with overall usage billed directly to the property owner. Submeters are typically installed by the customer to provide data on energy usage by area, allowing the customer to pinpoint buildings or spaces that use the most energy and need attention. Submeters can be used for billing, an example being an apartment building where tenants are billed individually for their electricity use.

Submeters typically have a couple of components to their deployment. One is a current transformer or "CT." The CTs are monitoring devices that look like a donut and lock on to individual electrical circuits at an electrical panel. The CT senses and gathers data on voltage, wattage and amperage on the circuit in real time or near real time. Multiple CTs will connect to a processor or a server/controller with the processor having a connection to an IP or BAS control network.

Typical communications protocols are Modbus TCP, SNMP MIB over an Ethernet connection, Modbus RTU over an RS-232 connection, and BACnet over an RS-485 network. Submeters are generally meant to provide information on energy consumption based on kilowatt hours (kWh).

Metering and submetering allow customers to make changes in their operations that reduce energy consumption. For example, if the data provided by a meter show that a large electric load exists when a building is unoccupied, there may be equipment running that needs to be shut off. The data may also show that shifting certain equipment schedules will reduce charges from utility companies by eliminating demands during peak periods.

Energy use indices (EUIs), such as kilowatt-hours used per square foot (kWh/SF), can be compared among similar buildings to determine if a building is using more energy than it should. EUIs can also be compared to previous data to see if the building's energy usage has increased. Since equipment efficiency tends to decrease over time, observations in energy usage can indicate when equipment is in need of service or replacement.

Smart Power Strips

Standby power or "vampire power" is electrical power consumed from electronic devices turned off or that are in standby mode. Studies in the United States, Britain, France, the Netherlands, Australia, and Japan show that the consumption of electrical power due to standby power is between 7 to 13%. The U.S. Department of Energy has stated that in the average home, 75% of the electricity used to power home electronics is consumed while the products are turned off.

This power consumption includes devices such as printers, cell phone chargers, DVD players, copiers, televisions, fax machines, and so on. It tends to be concentrated in areas such as offices and media centers. For example, the typical desk at one's office, whether at work or at home, has a personal computer, a monitor, probably a printer, a scanner, a VoIP telephone, a charger for a cell phone and so on. In the past, the only management choice was to unplug the devices, a move that may not be practical and/or safe.

Typically these devices are fed by a power strip with some surge protection. Current power strips have gotten smarter by incorporating microprocessors, thus allowing the strip to sense the electrical current and to monitor and manage the plug load. Some outlets on the power strip may be able to turn devices on or off by sensing whether the device is in use or in a prolonged idle state. The outlets can be turned off within a user-defined set time after the device goes idle. Other power strips are triggered by the personal computer; that is, the personal computer is plugged into a "control plug" on the strip and the power strip will shut down the peripheral devices when the personal computer is shut down.

Smart power strips can incorporate meters to provide information to users regarding energy consumption. Some go as far as providing energy costs and power quality, including voltage, line frequency, and power factor.

Smart power strips have also moved into data centers where rack-mounted strips feed servers, network switches, and other equipment. These power strips have an IP Ethernet port allowing managers to monitor, manage and reboot equipment intelligently.

POE

POE devices can be telephones, wireless access points, cameras, paging speakers, card readers and so on. Several major IT manufacturers have developed software to manage power to the devices, either turning them off and on or dimming the power to these devices, much like a lighting control system.

The power-over-Ethernet management software essentially enables and disables ports on a network switch. The result is a reduction in peak energy demand for IT networks and the flexibility for network managers to set different power consumption for various IT devices. POE management software is typically a module in a larger suite of network management tools with capabilities to scale from one network to an enterprise. Since the devices that are monitored and managed are already on an IT network, the monitoring of the devices is done with SNMP tools to evaluate any device defined by a Management Information Base (MIB).

Chapter 6

Access Control Systems

Overview

The vital importance of access control systems has increased along with security for buildings. The basic or typical building access control system operates so that a person presents a card to a card reader for a particular door and based on the information on the card and the system parameters for that person, door, and facility, the system either unlocks the door to allow the person to pass through or refuses entry. Similar operations apply to other areas in a facility where access needs to be controlled such as parking gates and elevators.

Doi:10.1016/B978-1-85617-653-8.00006-5

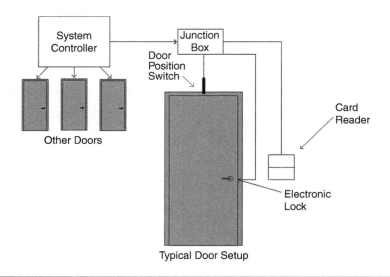

Figure 6.1   Access control deployment at a typical door.

The access control system is also important for life safety and is generally interfaced to the fire alarm system to facilitate building egress during evacuation. Access control systems must interface or integrate with several other smart building systems (video surveillance, HVAC, and others) as well as share data with business systems such as human resources and time and attendance. These systems have also become more important in providing information on building occupancy to other building systems such as lighting and HVAC systems where the demand for lighting or HVAC may be dependent on or coordinated with building occupancy.

The basic components of an access control system (Fig. 6.1) are as follows:

- A central host computer or server
- Control panels or system controllers connected to the host computer
- Peripheral devices such as card readers, door contacts, sirens, and sensors connected to the control panels

Server or Host Computer

The server or host computer houses the operating parameters and the database for the access control system. The host computer is networked and communicates with the control panels which collect data on events and alarms from peripheral devices.

The connection and communications between the control panels and the host computer generally occur using one of two methods. One is through an RS-232 or RS-485 communication from the host computer to the control panels. Typically this is accomplished by looping or "daisy chaining" a connection from control panel to control panel with all control panels accessing the host computer through the first control panel on the loop. The second and emerging technique is to have the host computer and the control panels on a network using IP protocols and standard structured cabling.

The user database for the access control system has profiles or credentials for every authorized user relating to levels and parameters. This database can be centralized in the host computer or distributed among the control panels with the host computer still having the complete database. This distributed network architecture is used to "push" the system decision making out to the local control panel, thus reducing communications traffic to the host computer.

For instance, when a person arrives at a door and presents an access card with encoded credential information to a card reader that information is then passed to the local control panel. In a centralized database system, the control panel passes the information on to the host computer with additional information regarding the location of the door and the time the card was presented. The host computer then verifies the information and compares the access level to the door location and the time of day. Upon verification, the host computer sends a command to open the door to the control panel and the person is allowed access to the building (Fig. 6.2).

In distributed database architecture a control panel with onboard memory contains the database for the doors and locations that it monitors. The control panel verifies the credential information without communicating to the host computer. The control panel later sends all transaction data to the host computer for system archiving. This distributed method can operate even if communication with the host is lost, including storing and buffering transaction data for sending to the host computer when communication is restored.

In addition to the database controlling cards and access levels the host computer is capable of producing administrative reports and creating users' badges. The badging system at the host computer typically includes a camera, backdrop and badging printer. Badging stations may be established in an organization's human resources or security department; anywhere employees, visitors, students, and others are entering or exiting the organization.

The software on the host system has features to facilitate the security operations. It may have the ability to supply an automatic notification, allowing security personnel to be notified via a PDA or email when a particular person

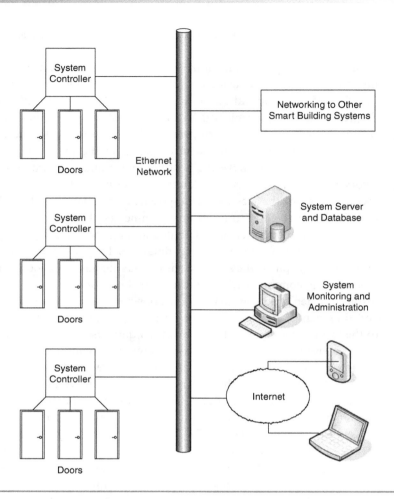

Figure 6.2 Access control system.

is attempting or has been granted access to a facility. Or, if a guard is stationed at the entrance to a building, the guard may use a personal computer that automatically displays information and an identification picture when a person enters the building using an access card, allowing the guard to compare and confirm that the person using the card is indeed the authorized card holder.

Control Panels

Control panels are usually enclosed printed circuit boards with connections to all peripheral devices in their area. These peripheral devices may include door hardware (such as a card reader, door position switch, or door strike) and other inputs and relays as required. The control panel manages the peripheral devices and communicates between the host computer and the peripheral devices. The control panels have the following functions:

- Consolidation of all connections to peripheral devices
- Provision of power, as needed, to peripheral devices
- Management of peripherals when communications to the host computer are absent or when acting in a distributed manner

Although control panels use a standard network connection and are able to communicate with the host computer or other control panels over the network, the signal methods of a control panel may be proprietary.

Most peripheral devices connected to a control panel do not have a lot of information to communicate. The device is either "on" or "off" or "open" or "closed." These devices (door contacts, requests to exit, door lock, auxiliary outputs and inputs) may communicate with a simple dry contact. For example, when the system detects a "short" across the contact, which indicates an event (e.g., the door is open), the system can take the required action.

The connection between the control panel and the card reader is more complex, entailing more information flow. The card reader typically uses the "Wiegand" wiring standard to communicate this information. The origins of the Wiegand communication protocol derive from card readers in the 1980s that used the "Wiegand effect" for sensing data. One of the advantages of the Wiegand standard is that it can be used for long cable runs.

The control panel communicates its status and recent access transactions to the host computer. The host computer or access server initially configures the control panel, may continuously synchronize the card holder database, and performs monitoring, management and administrative functions.

Peripheral Devices

Basic door-related peripheral devices are described in the following subsections.

Door Contacts

Door contacts or door position switches are nothing more than electromagnetic connections monitoring whether the door is open or closed.

REQUEST-TO-EXIT

The request-to-exit (RTE) device is inside a controlled door and detects a person approaching the door who wants to exit, and allows the person to exit without setting off an alarm.

ELECTRIFIED DOOR HARDWARE

These are components within the door or doorframe allowing the door to be automatically locked or unlocked.

CARD READERS

Differences among various access card readers are related to the technology and the interface with the person requesting access to a facility. Magstripe, swipe card, and insertion card readers use the same technology. Both proximity and MIFARE readers are "contactless" technology, operating at different frequencies, with MIFARE considered a "smart card" technology. Biometric readers use biological information to verify the identity of a person, and involve such measures as fingerprint scans, face scans, retina scans, iris scans, hand structure, voice identification, and other methods.

Other input devices that may be monitored or supervised by the control panel include:

- Push button—Primarily used for Americans with Disabilities Act (ADA) applications
- Panic button—Used to notify security of an emergency
- Glass breaks sensors—Notifies security of a breach to windows
- Motion detectors—Typically used in hallways
- Key pad—Key pads receive a code from a person and upon verification allow access

The host computer or control panel also has output relays that allow the access control system to interface with other smart building systems (like lighting control or HVAC) or a local or remote annunciator.

While the cabling for the door hardware (door strike, card reader, door contact, and request for exit) is typically not part of a structured cable system (typically two to four shielded twisted pairs, with cable gauges varying from

16 to 22), access control systems have embraced structured cabling and Ethernet for system controllers and the centralized server.

In addition, the requirement for the system to share databases with business systems has pushed the marketplace into offering databases that are compliant with SQL and ODBC standards.

IP POE-Powered Access Control Systems

Access control systems that incorporate IP and power-over-Ethernet (POE) components have been introduced to the marketplace in recent years. The evolution of access control systems to structured cabling, IP protocols, and POE was inevitable; a similar evolution took place in telephone and video surveillance systems. This transformation leverages the existing IT infrastructure, eliminates the need for local power, consolidates and saves labor costs for cable installation, reduces the time to install system devices, is more scalable and provides a large base of management tools and support. The move is subtly but surely changing the design and deployment of access control systems (Fig. 6.3).

In IP-based systems there is a direct network connection to either the door controller or the card reader with the network connection providing low-voltage power. If the network connection is to the door controller, POE powers the door contact, the lock, and the card reader. The door controller is really a gateway at the edge of the IP network converting the data and protocols from the door devices to an IP format. It has characteristics of typical door controllers such as the capability to buffer events or cache access credentials and features that address reliability and network performance issues.

If the IP access control system does not use a door controller, the network is connected directly to the card reader and the card reader powers the door contact and the lock. This approach not only eliminates the separate power supplies but also the door controller (Fig. 6.4).

POE POWER ISSUES

Some of the benefits and design issues with IP access control involve the elimination of high-voltage power and the use of POE. They are as follows.

Power Backup

In a traditional system backup power to door devices involves batteries and/or high-voltage circuits above each door on emergency power. In an IP-access control system if the network switches in the telecom rooms powering the IP

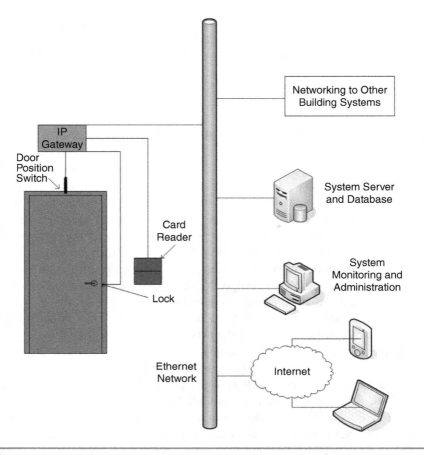

Figure 6.3   IP access control.

access control system are on UPS and emergency power, which they generally are, backup power is provided in a more centralized, less costly, more manageable way. With software applications residing on a network that can monitor and manage POE devices, deployment of an IP access control system provides more data, information, and manageability of the power than a traditional access control.

Power Sharing

A critical system design issue is selecting the right "power source equipment" which is either a network switch with POE capabilities or a device called a midspan that is used with a non-POE network switch to inject power into the communication connection. The issue is really with POE network switches

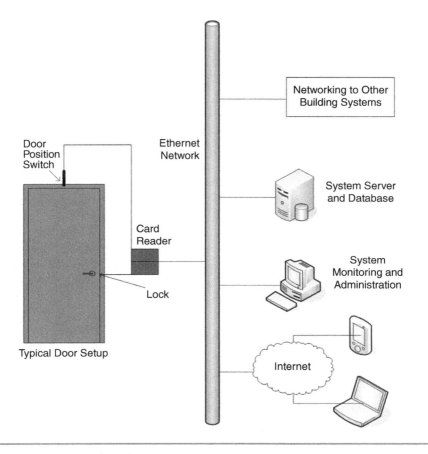

Figure 6.4 IP card reader.

that "power share," a characteristic of older POE network switches. These are switches that do not have the power capacity to supply full POE power to every port on the switch so they "share" the power among the switch ports.

The switch manufacturers figure that not every port will be needed for POE capabilities. Here's an example. Assume that you have a 24-port POE network switch with a 200 W power supply. The switch uses 50 W to power itself and has 150 W to provide POE to ports. With POE (IEEE 802.3af) at 15.4 W of power, only nine of the 24 ports on the switch could be provided POE. The potential risk is that connecting an access control device to a network switch using power sharing is lack of power to the access control devices. Ensuring that the networking equipment for an access control system will provide maximum POE power to every port is very important.

Power for Locks

Although each of the devices at a door (lock, card reader, request to exit sensor, door contact, etc.) has different power requirements, the lock is typically the most demanding. A lock requiring 600 mA at 12 VDC or 7.2 W essentially uses half the POE power. Locks are dumb devices; there are no POE or even IP locks. A mag lock, for example, simply gets a power feed and locks the door when there is power, and stops locking when the power is cut off.

The issue is really providing the power requirements for all the door devices that are within the current POE standard of 15.4 W. The yet-to-be-ratified POE Plus standard will resolve this power "budgeting" and provide 30 W of power.

Keep in mind that mag locks are inherently "fail-safe" devices (when they lose power, they unlock). As such, having a POE-powered reader feeding a mag lock could create a security issue since anyone who knows that the lock is fed by the card reader and wants to get through the door can just destroy the reader to gain access. In some cases it might be a better idea to use a fail-secure "mag" strike in order to prevent this, but such a method also has life safety considerations.

IP AND POE BENEFITS

Advantages of the IP- and POE-access control system approach follow:

Scalability—Instead of dealing with network or building controllers that may scale at four to eight doors at a time, the IP solution scales per door.

Reliability—The loss of one traditional network or building controller may take out four to eight doors; the loss of a network connection will take out one door in the IP approach.

Cost—There are considerable savings in infrastructure cost with IP systems. You eliminate the junction box above the door and the high-voltage power to it, network or building controllers, and the local battery packs. The cable contractor handling the IT network or video surveillance system can now also use cable for access control; thus, there are fewer contractors to coordinate and potential savings on cabling, labor, and cable pathways.

Integration—The IP solution is standards based; the cable and communication protocols are standards based, and hopefully specified to a standards-based database, all of which are integrated foundations at physical, networking, and application levels.

International Applications—POE is an international standard being marketed and deployed worldwide, allowing manufacturers to avoid supplying different power supplies by country, and thus eliminating the need for POE installers to worry about diverse equipment and power cords.

People Counters

People counters do not control access into buildings or spaces but do provide valuable information. People counters perform the simple act of counting people entering or exiting buildings or spaces within a building. Knowing how many people have entered a building or a space can be directly related to security, financial, or management objectives (Fig. 6.5).

DEVICES

People-counting devices used today are based on thermal or video imaging technology.

Thermal imaging devices, the more traditional approach, sense a person's body heat compared to the background thermal image. The people being

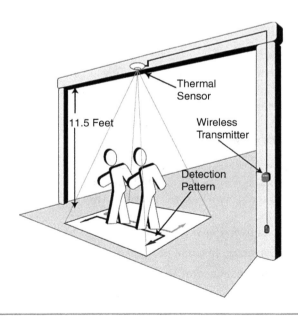

Figure 6.5 People counting using thermal sensing.

counted have no sensation from the counting device and their identity is not registered (that's left to the video surveillance and access control systems). The thermal imaging is not affected by lighting or other conditions. Devices are typically mounted 8 to 15 feet high.

The devices are relatively low cost and can be easily installed. Each device contains a sensor, imaging optics, a signal processor, and some type of networking interface to connect to other devices or the main administration terminal. Multiple devices can be used for larger-width door openings and can be programmed to avoid double-counting, essentially acting as a single device.

Video imaging typically uses small cameras with an auto-iris lens and video analytics to count and differentiate among people entering and exiting a building or space. People-counting capability as an add-on module of a video surveillance system is not unusual.

Some of the actionable information that can be created from people counters includes the following:

Access Control

One of the typical issues with access control systems is "tailgating" or "piggybacking." Tailgating is when someone with a valid access card enters a building but is immediately followed by an unauthorized person before the door closes. A people-counting system would accurately detect "tailgaters," helping prevent unauthorized access, resulting in a higher level of security and less theft or vandalism.

Life Safety

Data on the number of people leaving buildings in the event of an emergency evacuation and the number of people who had entered the building are valuable information for first emergency responders.

Financial Metrics of Retail Stores

People-counting systems probably originated in the retail sector and large venues (theme parks, stadiums, etc.). Retail stores can easily correlate foot traffic into the store with the store's business metrics, answering questions such as:

- What's the conversion rate of foot traffic to buyers?
- What's the average revenue per person entering the store?
- What are the seasonal traffic patterns?
- How do similar stores in different locations compare?
- How can the number of sales staff be optimized to the number and patterns of people entering a store?

■ What's the impact of a sales event on foot traffic?

■ What's the impact of new opening hours?

Tracking Audiences for Advertising and Self-Service Facilities

People counters can provide information on how many people view advertisements or information on digital signage or electronic tickers and activity on self-service devices such as kiosks. Data on the number of viewers and users of these devices are useful for evaluating the advertising and/or the placement of equipment within a building.

HVAC Control

People counters are useful for optimizing the air quality and energy use based on actual occupancy of a building or space. Actual people counts can be used at the beginning of the workday to start up and ramp up the HVAC system properly. People counting can be used upstream as well as downstream in the ventilation of certain spaces.

For example, an advanced HVAC control approach is CO_2 demand control ventilation (DCV), which is best used in large areas, open office spaces, theaters, assembly areas, ballrooms, and so on. A CO_2 sensor is deployed to optimize the use of outdoor air and the energy required to condition the air.

The CO_2 sensor is really a "people counter." People exhale CO_2, and the CO_2 sensor reflects the occupancy of the space by the level of CO_2. But there are limits to sensing occupancy through CO_2 sensors, and at times such sensors are unreliable and provide poor estimates of occupancy. People-counting technology with accuracy rates of 95% provides more reliable and accurate estimates of occupancy.

Space Planning

Fifty percent of the space in a typical office building is unoccupied during the business day. People counters can provide information on how buildings, floors, and spaces are actually being used, thus allowing a company to optimize the allocation of employees to spaces and to evaluate options such as "hoteling" or "hot-desking."

Staff Management

People counters provide data that management can use to better understand a building's traffic patterns, event attendance, seasonal impact, time-of-day differences, and so on. With such data in hand, management can then determine appropriate staffing and security levels to ensure optimal services and sales.

People Management in Entertainment and Other Large Venues

An important part of managing a sports, entertainment, church or mosque, public transportation system or theme park venue is people management; that is, managing large volumes of people in and around the building to ensure that they enjoy/use the venue and do so safely.

The people-counting system can be used to properly queue customers for access to rides, detect potentially dangerous situations, validate door entry numbers and optimize the placement of retail and food spaces within the venue. Occupancy levels in large buildings must be monitored so that alarms are issued when building capacity is exceeded. Passenger counts on public transportation systems are used to maintain and enhance efficient and responsive route planning.

Counting people entering and exiting a building or facility consists of very basic data that every facility manager, property manager and owner should have. Installing a system is not difficult or expensive and the data generated are easily understandable and can be used in many ways. Such data provide key components to optimize operational planning and analyze business trends.

Chapter 7

Video Surveillance Systems

Overview

Video surveillance systems, also known as closed-circuit television systems (CCTV), are part of a facility's larger security and life safety plan. The larger plan may include physical and operational aspects of security as well as other security or life safety systems such as access control and intrusion detection.

It is critical to note that the deployment of video surveillance systems must take into consideration important legal aspects, mainly a person's right to privacy and the presumption of security. For example, one should not place a camera in an area where an individual should expect privacy, nor should one deploy "dummy" cameras so as to provide the perception of security monitoring, when, in fact, none exists.

Video surveillance systems have for decades been based on analog technology. The change and evolution to digital technology follows a trend similar in the broader electronics marketplace.

Major Functions

Video surveillance systems basically perform five functions:

- A video surveillance camera captures a picture or video image of an asset or access to that asset that needs to be secured or protected from theft, tampering or destruction.
- The video is transmitted back to a security control center.
- The video is processed.
- The video is recorded.
- The video is viewed on a monitor (Fig. 7.1).

VIDEO CAPTURE

The picture or video is captured via the camera and lens assembly. Surveillance cameras are traditionally either fixed cameras, which provide a single view, or pan/tilt/zoom (PTZ) cameras, which provide several views and can be controlled

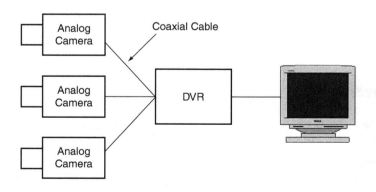

Figure 7.1 Typical video surveillance system.

remotely from a security control center. A third type of camera, called panoramic, is emerging. Panoramic cameras provide a 360-degree view of an area.

The panning and tilting of a PTZ camera are accomplished by the camera mount, not the camera itself. The one drawback of a PTZ camera is that while it views a specific area within its range, some event could take place in another part of its panning area and go unnoticed. To compensate, PTZ cameras can pan in a predefined pattern or can be interfaced with other systems, such as access control, so that they automatically pan to a specific area when triggered.

Cameras come in a wide variety of resolutions and some have advanced features to control and adjust lighting. Based on the security required, placement of security cameras must consider the camera, lens, lighting, and power.

Variable lenses are the most commonly used, allowing for adjustment in the field. Zoom lenses can be adjusted remotely from the security control center. Color cameras are typically used in most video surveillance systems although black-and-white cameras are actually better in low light situations. Combination ("night/day") cameras are available for exterior use where lighting varies regularly. Camera mounts and housings for cameras are available for a variety of placements and aesthetic tastes. Housings are also available for special environments, such as prisons or laboratories.

The true revolution in cameras has been the digital surveillance camera coupled with an IP network connection. A networked camera digitizes the video (typically into a MPEG or H.264 or Motion JPEG), compresses the digitized video, creates a data packet of the video and transmits it over a standard data network. The cameras may have the capability to store video, buffer it if network traffic is heavy or process the video. In addition to the digital video output, the network connections can transmit PTZ signaling, audio and other control and management commands. The cameras can also be powered from a central location utilizing POE.

VIDEO TRANSMISSION

Transmission of the video signal captured from a surveillance camera to the security control center has typically occurred through coaxial cable, the traditional cable for analog video. With changes in the technology more installations are using unshielded, twisted-pair copper cable, fiber optic cable and wireless solutions. Unshielded twisted pair is even being used with analog cameras, with baluns (an interface between *bal*anced signals and *un*balanced signals) or a manufacturer's proprietary technology, which may allow signaling over long distances.

With IP cameras transmission is accomplished through unshielded, twisted-pair cabling as part of a structured telecommunications cabling system. Fiber optic cable is utilized for exceptionally long cable runs or for exterior cameras where lighting protection is a concern. The distance between the camera and the headend equipment, as well as cost, security of signal, and resolution, may be considered in selecting the physical transmission media.

Wireless transmission can be used for cameras where cable is impractical or costly. Wireless can be deployed rapidly but it may require power and line of sight between locations; it may also be susceptible to interference. Wireless technologies include "Wi-Fi," infrared, microwave and free-span optic (FSO) systems.

VIDEO PROCESSING

Video processing codes or encrypts video signals from surveillance cameras. The processing allows multiple cameras to be displayed on a single monitor or multiple camera views to be cycled on a monitor. Two products, both of which are remnants of analog video systems, have typically been used to accomplish this: a cross-point matrix and multiplexer.

The basic functionality of the older cross-point matrix is to switch camera feeds to different outputs. It takes the feeds or inputs from the video surveillance cameras and "binds" any input to any output or to multiple outputs.

Older video multiplexer functionality can be summarized as taking multiple video feeds and combining them into a single video feed. This allows multiple cameras to be viewed on a single monitor. Multiplexing does so by lowering the resolution of each video feed and rescaling it to a screen, typically showing the views of 4, 9, or 16 cameras.

In a digital system with IP cameras and a network server replicating a cross-matrix and multiplexer functionality is significantly enhanced through a software-dependent approach rather than the hardware-dependent approach of traditional analog systems.

RECORDING

One of the first components of video surveillance systems to go digital was the digital video recorder (DVR) introduced to replace the older, tape-based video cassette recorder (VCR). The VCR worked with analog video and typically

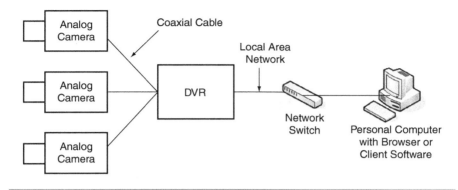

Figure 7.2 Video surveillance—networked (DVR)

sat behind a multiplexer, this made multiplexing several cameras onto a single video cassette tape possible (Fig. 7.2).

DVRs were first introduced in traditional analog systems. DVRs can digitize video inputs from analog video surveillance cameras and also have the functionality of a multiplexer. Hard disks are the recording media for DVRs, thus eliminating the need to consistently change out or store tapes.

DVRs evolved further to include an Ethernet network connection, thus allowing the digital video to be transmitted over a data network and opening up the possibility of viewing video remotely through a web browser. Digital recording brought new functionality to video surveillance such as the capability to detect motion in a picture, to record at different frame rates based on the detection of motion, view video while recording video and more.

The next digital evolution of video recording occurred when the DVR was replaced with a video server, that is, a data network server with video management software. In this arrangement analog or digital cameras connect to the server and the server connects to the network. The video server uses standard data network equipment and becomes the centerpiece of a video surveillance system. Because the server is on a network it opens up a wide variety of functions for recording, storing, viewing, and administering the system, either on the network or off site (Fig. 7.3).

A major advantage of the video server is that the functionality of the system is derived from software rather than hardware as was the case with older analog video surveillance systems. For example, an administrator or an authorized user can determine and set the number of inputs, the resolution required and so forth for viewing multiple cameras or video on a monitor or screen.

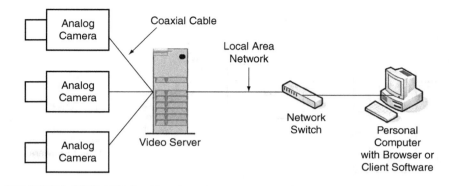

Figure 7.3 Video surveillance with video server.

MONITORING

Monitoring is categorized as the viewing and analyzing of video from the surveillance cameras and, if warranted, taking other security actions based on the analysis. Traditionally this has been accomplished in a security control room with banks of monitors and security personnel. The viewing can be live video feed or archived recorded video. CRT technologies, which have traditionally been used in these control room environments, are being replaced with LCD and plasma monitors (Fig. 7.4).

More importantly, the digital transformation in video surveillance has provided a variety of ways for monitoring to take place. Authorized users with an Internet connection and a desktop, laptop or PDA can access the video

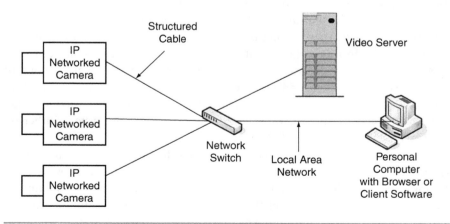

Figure 7.4 Full digital video surveillance system.

surveillance system remotely and view video. Or, the system can be set up so that when an event is detected on a camera an image is transmitted to a local or remote device for specific security personnel.

In addition, as the technology systems in a building are integrated and converged, separate control rooms, such as those for security or data network control can be integrated into a more comprehensive network operations center where all smart building systems are monitored and administered. This convergence of organizations mirrors the convergence of system technologies. Operation centers that monitor multiple systems or buildings have proved to be very effective in reducing operating costs.

Analytics

Video analytics involves monitoring and analyzing digital video for specific objects, facial recognition, behavior, occupancy and so on. Although the software technique is not 100% accurate and may give false alarms, it is very useful in garnering additional data from the video. Video analytics are used to identify occupancy, license plates, faces, and the density of people or cars.

People-counting systems, previously discussed in Chapter 6, can use video analytics to count people. In this case a camera is pointed down to an area and identifies which way people are moving. Some airport security systems use something similar in identifying counter flow, that is, objects or persons moving against the flow. Another application is for security staff to "draw" a line on a fixed video image, such as an area where people or vehicles are prohibited, and then receive an alarm when motion is detected in the prohibited area.

A video analytics software application can be a separate standalone system from the primary video surveillance management software or integrated into the software as a module with the integrated approach being a preferred method. In addition, analytics can occur at the network camera or at the central management software. Analytics on live video should occur at the camera since it has a dedicated processor and uses network bandwidth.

Analytics on archived or recorded video is best accomplished at the central server where video can be searched for different actions or objects. Analytics help security staff locate events in the video, a task that takes considerably longer when manually performed by the security staff.

IP-Based Video Surveillance Systems

Digital video surveillance systems using standard structured cable and IP and Ethernet protocols allow the system to easily become part of a smart building.

They also demonstrate that digital systems provide increases in system flexibility, functionality, and scalability. Some of the advantages as well as the deployment issues for IP-Based video surveillance systems follow.

CONCERNS

The move to IP-based systems for video surveillance and access control requires substantial coordination between an organization's security department and the IT department. Security personnel may lack knowledge about IP addresses, domain name servers (DNSs), port assignments and so on.

The quantity and resolution of video surveillance cameras may affect other network traffic and the overall quality of service of the network. Also, while IP-based systems means open and interoperable systems, equipment may use different video encoding methods or management software solutions may be available for only certain types of cameras.

ADVANTAGES

IP-based video surveillance leverages existing IT infrastructure and contributes to lower cable installation costs as the number of cable contractors is reduced and cable pathways can be shared. It reduces the quantity of cable to PTZ cameras. It also allows for consolidation of equipment room space and environmental support for the space. The IT infrastructure allows for improved network security, remote system access, remote notification of events and alarms and use of POE. Most importantly, it allows for the integration of the video surveillance systems into other building technology systems allowing for greater functionality between the systems.

IP-based systems have greater flexibility in image format and provide greater image resolution and increased flexibility in configuring frame rates and resolution for each camera. The capabilities to use wireless connectivity, digitally zoom into an image and incorporate older analog cameras into a system through encoders are major benefits of the digital technology.

Chapter 8

Video, IPTV, and Digital Signage Systems

Overview

Video programming distribution is related to but different from audio visual systems. The best example of enterprise video distribution is cable television where a service provider is distributing video programming to its customers. Within a building the video distribution network is distributing video and informational content. The content may be re-broadcast content from a cable television company but may also be advertisements, public safety information, archived educational programs, live video feeds, entertainment, and energy information.

Doi:10.1016/B978-1-85617-653-8.00008-9

Video distribution varies by building type. For residential buildings such as houses, apartments, condominiums, and dormitories video distribution is fairly dense because most television viewing takes place at home. In most commercial and government buildings video distribution is primarily in common areas such as building entrances, cafeterias, meeting rooms, elevators, assembly rooms, classrooms, and so forth.

Traditional Video Distribution

Most video within existing buildings is currently being distributed using a technology developed decades ago, commonly referred to as CATV (community antenna television) or RF (radio frequency). The technology takes multiple analog video signals and "modulates" or places them on different radio frequencies carried by a coaxial cable. Many large cable television service providers and others have moved away from the use of coaxial cable to some hybrid of fiber optic cable and coaxial cable or, in some areas, just fiber optic cable, to transmit the video RF signal (Fig. 8.1).

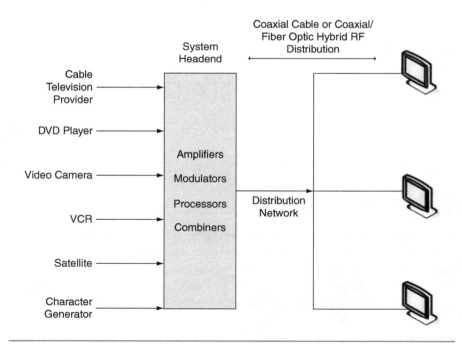

Figure 8.1 Traditional video distribution.

Video distribution in many buildings is simply a retransmission of programming provided by the local cable company. More sophisticated systems may selectively filter or block some of the programming from a local cable company or insert local sources of video from a DVD, camera, data network or character generator.

Video distribution, like other technology systems in a building has evolved and is moving to digital formats, standard cable infrastructure (unshielded twisted-pair copper and fiber optic cabling) and the IP protocol. The result is IP television or IPTV. In many ways video distribution is following the evolutionary technical path that voice, wireless, video surveillance and access control have taken.

Video Display and Viewing

In 1941 the U.S. National Television Standards Committee (NTSC) proposed the original analog television format for black and white television, following in 1950 with a backward compatible standard for color television. Many of us still watch television in the 1950 standard.

Analog video is a group of still images or frames. Thirty of these frames are broadcast each second giving viewers a visual sense of motion. The quality of the resolution of any video transmission is determined by the number of horizontal lines displayed on the screen and the number of tiny dots, or pixels, that emit either red, blue, or green light to create the picture we see on the screen. The total number of pixels is the measure of the quality or resolution of the picture.

The 1941 analog standard called for 480 horizontal lines of resolution (additional lines are reserved for synchronization and information such as captioning), 720 pixels per horizontal line, and a technique called interlacing. Interlacing means that only half of the horizontal lines are used in creating each frame; that is, even-number lines are broadcast in one frame and odd number horizontal lines are broadcast in the next frame. The total resolution of a screen can be calculated by multiplying the number of pixels in each line by the number of lines on the screen (720 pixels per horizontal line for 480 lines = 345,600 total pixels).

There are currently 18 different digital television formats set by the Advanced Television System Committee (ATSC) standard. These are formats for televisions and displays that can be grouped as follows:

Standard Definition (SDTV)—This is basically the analog television standard delivered as a digital signal with three variations of the formats dealing with the number of pixels, the shape of the pixels, and use of interlacing.

Enhanced Definition (EDTV)—EDTV is like SDTV with the main difference being that it uses progressive scanning rather than interlacing. There are nine variations of EDTV dealing with the number and shape of pixels, as well as the aspect ratio.

High Definition (HDTV)—HDTV offers the highest-quality resolution and uses a 16:9 widescreen aspect ratio only. HDTV has six variations addressing lines of resolution, number of pixels, frame rates, and interlacing.

Digital Video Transmitted via a Data Network

If one were to digitize the traditional analog video signal, assuming that each pixel needed 3 bytes of information to indicate color, the digital signal would need a transmission rate of over 8 Mbps, a high-bandwidth network requirement that is still beyond the capability of the vast majority of data networks. To reduce the size of digital video several techniques were deployed; the number of frames per second was decreased, the sizes of screens were reduced, transmission for pixels that did not change frame to frame were eliminated, compression techniques were used and so forth.

The result was the development of hardware or software codecs (*co*ding and *dec*oding devices) that compressed analog video and decompressed digital video signals. Codecs were developed around a series of compression standards such as the H320 and H323 standards.

Video can be digitized in several formats but typically uses an MPEG (Motion Pictures Expert Group) standard for the encoding and compression of digital multimedia content. MPEG-2, or its latest incarnation, H.264, provides compression support for the TV-quality transmission of digital video. Other standards address channel change signaling and video on demand.

Delivery of video over a standard data network can be performed in several modes:

■ Multicast Mode—programming can be sent to multiple devices at the same time

■ Unicast Mode—programming can be sent to just one device

■ Broadcast Mode—programming can be sent to all open devices at the same time

A typical deployment of video distribution (Fig. 8.2) in a smart building involves:

■ Digital television transmitted over a TCP/IP-based data network

■ Digital video distribution that transmits and receives MPEG video streams

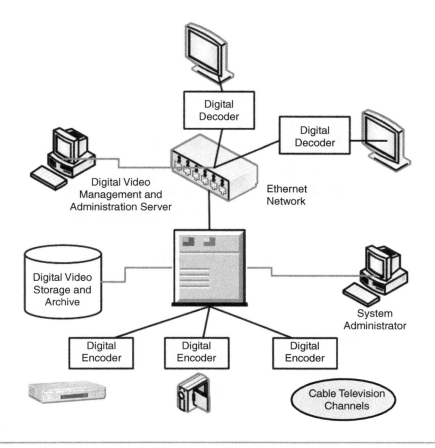

Figure 8.2 Digital video distribution.

- Distribution and transmission of digital video in a broadcast, multicast or unicast mode
- A media retrieval system, allowing authorized users to schedule and stream video programming on demand
- An extensive management and reporting system
- Inputs to the digital video distribution system, including live or recorded video, encoded analog video, cameras and character generators

Applications

Beyond the basic video distribution lie many forms of enhanced video distribution methods and applications. For example, video distribution can be point-to-point or point-to-multipoint, one-way (unidirectional, much like the delivery method of cable television into a house) or two-way (bidirectional, where video signals are sent upstream to be retransmitted downstream). Some of the applications are summarized in following sections.

MEDIA RETRIEVAL

In this application video programming is stored in digital or analog format and a system administrator or authorized individual users can schedule, or "call up," in real time, specific video programming. For example, a mathematics teacher, possibly with the assistance of a librarian or media director, can call up a program entitled "Practical Applications of Algebra" for viewing by the teacher's 2:00 PM algebra class. A similar application, video on demand (VOD)—primarily used for movies—is also becoming more mainstream with the hospitality industry.

VIDEO CONFERENCING

Video conferencing technology addresses the cost and potential inefficiencies of live meetings. Its deployment differs by application. It can be personal video conferencing where participants use a personal computer with a web camera. In smart buildings the application comprises video conferencing equipment for a typical meeting room where displays and audio equipment provide for the exchange of video, content and audio. Most recently this type of video conferencing has evolved to "telepresence." Telepresence shows every meeting participant in true-to-life dimensions to the point that one can read the body language of others. Each connected telepresence location is equipped with a duplicated surrounding, such as similar table, chairs, wall coloring, and so on, and provides a sensation that each participant is seated around the same table. Regardless of the type of video conferencing deployment, the audio quality between locations is critical to its success—video can embellish and enrich the audio conversation but cannot replace it.

DISTANCE LEARNING

Distance learning can be described as a specialized form of video conferencing. Typically distance learning has an instructor in one location and a class or individual student at another location. The application may also be augmented with graphics transmissions. Classrooms are equipped with a series of speakers and microphones. The application can include subsystems, such as remote response systems, where students have a keypad with audio capabilities, thus facilitating communications between students and the instructor, and even testing.

LIVE FEEDS FROM VIDEO CAMERAS

In a building or campus environment a live camera feed from a CEO or principal's office may be distributed to employees or students gathered in meeting rooms, classrooms and common areas.

Digital Signage

Digital signage is a very compelling technology. The medium has "stickiness"; few people will not at the very least, glance at or pay attention to a plasma or LCD display. It's a communication system that's effective, immediate, and dynamic. It can be used in a variety of building types.

The market for digital signage now seems driven by the lower costs of flat panel displays and also by the opportunity for advertising revenues in some venues. Inasmuch as digital signage has come of age it offers important functionality in smart buildings. So how does digital signage work? And more importantly, what are its best uses in buildings?

DIGITAL SIGNAGE SYSTEMS

Digital signage is an application on IPTV systems that broadcasts content to a viewer (as opposed to static signs). The digital signage networks are scalable and can be used in one room, a whole building, or a metropolitan area and beyond (Fig. 8.3).

The basic components of a digital signage system are very straightforward. They include displays, media processor and controller, management workstation, and content as described in the following sections.

Displays

Video displays come in a wide variety of sizes and core technologies. Selection of a display is based on the dimensions of the viewing area, lighting, and the type of content and materials to be displayed. The most popular displays are plasma screens and liquid crystal displays (LCDs).

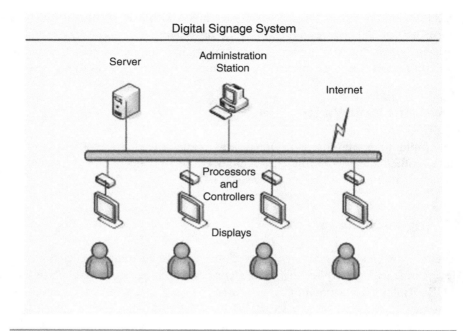

Figure 8.3 Digital signage systems.

Plasma screens are basically a network of chemical compounds called phosphors contained between two thin layers of glass that when excited by an electric pulse, produce colors, light, and a picture. The image of a plasma screen is bright, rich with color and has high resolution. Screens can be as large as 103 inches. Plasma screens are thin, lightweight, have a wide viewing angle and offer good images under normal room-lighting conditions. The downsides are that plasma can be pricey and also prone to "burn in."

LCDs use a fluorescent backlight to send light through its liquid crystal molecules. LCD monitors apply voltage to the pixels to adjust the darkness of the pixels, thus preventing the backlight from showing through. An LCD display can be as large as 100 inches. LCDs are very thin, lightweight, and have good color reproduction and sharpness. LCDs are typically less costly than plasma and may be more suitable for brightly lit areas.

The size and mounting of the monitors are both an architectural as well as a functional issue. The size of the monitor should be dictated by the distance at which people will be viewing the monitor. However, there are no rigid scientific rules on display sizing and viewing distance.

A general rule of thumb can be the anticipated viewing distance should be 2 to 5 times the width (not the diagonal measurement but the width) of the monitor. For example, a 2-foot-wide monitor is best viewed from 4 to 10 feet. This may vary for standard television signals (3 to 5 times) and high-definition television (2 to 4 times).

Some digital media systems can provide video media to displays for signage as well as for viewing via a personal computer. Touchscreens can also be used with systems if interaction by the viewer is desired, allowing the viewer to step through menus or select programming.

Media Processor and Controller

The media processors and controllers receive content from the main headend system, locally store the content, and then feed content to the display as directed. The controllers are networked via a local area network or the Internet to the headend servers and management workstation. The processor and controllers are managed by the networked, attached administration workstation but also may have options for local management of the controller.

Controllers will usually be provided with multiple graphic formats and screen resolution capabilities as to be compatible with a wide variety of displays. Some controllers have an option for wireless connectivity to the network.

Depending on the configuration, some systems may feed the displays directly from the headend and utilize a decoder at the display location to convert the digital video into the graphic format needed by the display.

Management Workstation

This is a PC connected to the network and loaded with the appropriate system software. The software allows the system administrator to create the content, schedule content distribution, configure which media processors and controllers or displays get what content, monitor the system and generate management reports on the systems.

Media Players and Servers

Excluding the live feeds for system display, the media server stores the content for the digital signage system in a digital format. This may include digital or MPEG video, content downloaded from the Internet, content created with PowerPoint or QuickTime, text, still images and so on. For a very small deployment a media player can drive one display and be updated via CD-ROM or USB flash drive. Larger systems with network connectivity allow digital signage network operators to either push content to many players at once or have each player pull content from a server as needed.

Content

Procuring and installing the digital signage equipment is the easy part. The harder part is continuously creating the content for the audience and purposes intended, evaluating the effectiveness of the content and creating and updating content. Digital signage systems require a commitment of resources to create and manage the content. A digital signage project should not be undertaken until one understands the ongoing needs of the system, some of which can be handled by a third party.

USING DIGITAL SIGNAGE

Digital signage can be used to inform, entertain, communicate, advertise, and enhance people's overall experience in the building. Digital signage uses will vary according to building type. A few examples are shown in Table 8-1.

Table 8-1   Uses of Digital Signage by Building Type

Education	Business
■ Class schedule ■ Campus television network ■ Registration and orientation information ■ Schools news and weather announcements ■ Local event calendars ■ Media retrieval for education ■ Distance learning	■ Employee recognition ■ Executive communications ■ Service and product training ■ Company news ■ Performance and quality control statistics ■ Announcements, holiday and special events information ■ Conference room scheduling
Government	**Healthcare**
■ Public notices ■ Emergency alerts and information ■ Weather and news ■ Meeting room schedules ■ Way finding ■ Communications to geographically dispersed government departments ■ Training	■ Way finding ■ Patient queue information ■ News and weather ■ Healthcare information for patients ■ Directories ■ Staff training ■ Communications to geographically dispersed healthcare organizations

Figure 8.4 Use of digital signage for life safety.

One of the most innovative uses of digital signage is for life safety, where a digital signage system augments the fire alarm and building automation systems. Digital signage is located near the traditional "exit" signs at egress points. In an emergency such as a fire alarm if a stairwell is not safe for evacuation, the digital display can provide a message such as "Smoke in Stairwell, Do Not Enter, Saferoom is Room 12." The system can allow the fire command center or emergency command that responds to the alarms the ability to control the signage and view the situation through an optional camera, thus improving evacuation and response (Fig. 8.4).

Deployments of digital signage systems are facilitated by existing data networks, shortening commissioning times. The uses and content of the systems are very flexible and can be tailored to organization needs. Video distribution will continue to penetrate, evolve, and be delivered through standard smart building infrastructure.

Chapter 9

Fire Alarm and Mass Notification Systems

Overview

Fire alarm systems are the primary life safety system for every building. Properly deployed, a fire alarm system reduces the probability of injury or loss of life and limits damage due to fire, smoke, heat and other factors. Because of their criticality, the codes, regulations and standards affecting the design and installation of fire alarm systems are wide ranging and detailed. Their design and installation must involve qualified, licensed, experienced professionals and more important, the coordination and approval of the local authority having jurisdiction (AHJ).

Doi:10.1016/B978-1-85617-653-8.00009-0

The two main organization addressing codes and standards for fire alarm systems are the National Fire Protection Association (NFPA) and Underwriters Laboratories (UL). NFPA 70, 72, and 101 address the National Electrical Code, National Fire Alarm Code, and Life Safety Code, respectively.

The National Fire Alarm Code addresses system design, location of devices, testing procedures, performance requirements, and maintenance procedures. The National Electrical Code covers the equipment and wiring of fire alarm systems. The Life Safety Code covers more than buildings per se as it identifies construction, protection, and occupancy features necessary for life safety.

Underwriters Laboratories is an independent product safety certification organization that tests products and writes safety standards for fire alarm system components such as control panels, smoke detecting heads, horns, and pull stations.

Even given their life safety nature fire alarm systems are starting to utilize IP-based functionality and should be integrated with other systems within a smart building. The integration to other systems plays a critical role in minimizing the effects of the fire. A fire alarm system will initiate communications to other building automation and security systems to facilitate evacuation from the building and containment of the fire.

Such systems include the following:

- Signaling the HVAC system to restrict and contain smoke, heat and fire through dampers and fans
- Using the access control system to clear a path for building evacuation by opening doors, unlocking secured doors, and releasing powered exterior doors
- Using the access control system to contain and prevent the spread of fire and smoke by closing interior doors
- Triggering emergency power for the fire alarm system and related systems operation, exit signs, and lighting for building exit routes
- "Capturing" the elevator and shutting down its operation

The fire alarm system must communicate with and control its system components and it must also communicate with offsite facilities and organizations such as the fire department and emergency services. The networking of the fire alarm system components, like other systems, is accomplished with a cable infrastructure and communication protocols.

The reliability of a fire alarm system is partially dependent on the system cabling. Both the National Electrical Code and NFPA have specific guidelines to ensure proper system operation. Most alarm system devices must be cabled so that there is a redundant wiring path from the control panel to ensure that the device will function in the event that the cabling is damaged. Survivability is critical.

Cabling between control panels may be standard structured cable, such as twisted pair or fiber optic, allowing the parts of the fire alarm system to use the same structured cable infrastructure that is used by other smart building systems.

The communications protocols used by major fire alarm systems manufacturers typically conform to a BAS protocol such as BACnet or LonTalk while some have introduced use of the IP protocol between major components (Fig. 9.1).

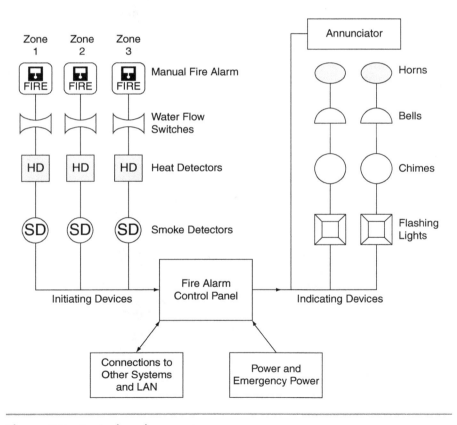

Figure 9.1 Basic fire alarm system.

Fire Alarm Control Panel

The heart of a fire alarm system is the primary fire alarm control panel (FACP). The FACP monitors system integrity and starts all sequences of operation for system detection, suppression, and notification. FACPs are typically microprocessor based with software for communications, processing and decision making.

FACPs gather data from system detection or indicating devices, process the data, and then act on the data if need be; managing system alarms and suppression systems as well as other building automation, security, and life safety systems. FACPs can also detect failures within the system that require repair and maintenance (Fig. 9.2).

Depending on the size of a building or the number of buildings on a campus, an overall system may have a single central control unit or may have distributed control units networked together to provide a unified and transparent system. Field or remote control panels can be networked to a central panel. The relationship between the central and field panels can be "peer-to-peer" or "master-slave." Field panels can be deployed for narrower, more specific functions such as simply supplying power to alarm devices or used as a remote annunciator.

System devices assigned to a zone can be connected to the FACP either through a direct connection, bus or loop topology. The NFPA designates the performance of these connections or circuits. Class A circuits must be able

Figure 9.2 Fire alarm control panel.

to transmit a signal even with an open or ground fault on the circuit while Class B circuits need not. The loop connection can be "fault-tolerant" thus allowing continual operation if there is a break in the cable or a device failure.

The connection of the FACP to the devices can be "addressable" or "supervised." Before devices were addressable an alphanumeric display at the control panel indicated which zone was affected. Addressable means that the FACP can communicate to a single device or a group of devices depending on the functions required and know the exact location of the device. Supervised devices are monitored to ensure that they are still connected to the FACP and are operational.

Regardless of whether devices are supervised or addressable the system is required to continuously monitor their status. For addressable devices monitoring is accomplished by polling individual devices. For nonaddressable devices monitoring is accomplished by sensing electric current. The FACP provides a small current which, if interrupted, indicates trouble such as a failed or missing device.

A control panel is fed by standard branch-circuit wiring and typically includes alphanumeric display and touchpad to provide information on the state of the system. This allows a technician to shun an alarm or reprogram or reset the systems. The panel may include batteries to provide power for its operation if AC power fails.

ANNUNCIATOR PANEL

A major component to the system is the annunciator panel attached to the FACP. The annunciator provides visual and audible indications that an alarm has been initiated as well as the location of the alarm. It may also identify the functions that could affect the fire and the building occupants in the area.

A basic annunciator panel may have an alphanumeric display, and switches to acknowledge and silence the alarm. More sophisticated annunciator panels are equipped with a monitor or include a personal computer with a GUI displaying floor plans of the building and the location of the alarm.

FIRE DETECTION

Fire consists of smoke, heat, and light. The system components that detect the fire and initiate an alarm monitor one or more of the fire's characteristics. The detection components of a fire alarm system are typically located in ceilings, HVAC ducts, mechanical and electrical areas and equipment rooms.

They include but are not limited to the following:

- Pull stations, in which a person sees a fire and pulls the fire alarm
- Thermal detectors, which sense a rise in temperature or the high temperature of a fire
- Smoke detectors, which sense vapors of small particles of carbon matter generated by burning, including in-duct smoke detectors
- Flame detectors, which sense radiation and visible light from a fire
- Sprinkler water–flow sensors
- Fire-gas detectors, which sense gases such as carbon dioxide and carbon monoxide
- Air-sampling fire detectors, which are the most sensitive type of detection available, and are used in high-value and critical environments such as churches, clean rooms, hospitals, museums and communications or network equipment rooms; this system typically uses tubes to continually draw air samples to a highly sensitive detector, and can detect a precombustion stage of a fire prior to any visible smoke or flame

SUPPRESSION SYSTEMS

Fire suppression systems include the following:

- Wet sprinkler systems, which may comprise various switches and flow-detection equipment that is monitored and managed
- Dry sprinkler systems, which may include pressure switches that are monitored and managed
- Fire suppression systems may also include monitoring and supervision equipment

NOTIFICATION DEVICES

Once a fire is detected a building's occupants must be notified to evacuate the building. Fire notification devices use audio, visual, or a combination signaling to notify occupants. These devices are typically DC-powered so that they can operate on backup batteries and they must also adhere to product compliance as determined by Underwriters Laboratories testing and listing guidelines for use as a fire alarm notifier.

Fire alarm notification devices include, but are not limited to, the following:

- Bells
- Chimes
- Horns
- Speakers
- Strobes, including strobe lights combined with other devices

Monitoring

There are two fire alarm system classes:

1. A protected premise system, which is a single building or campus of buildings under control of one owner and protected by a single system where the system is monitored locally or remotely by the owner

2. A supervised station system, which is much like a protected premise system except that the system is continuously monitored by a third-party security or central monitoring company; monitoring is addressed in the following ways:
   - *Local monitoring*—When activated, a local alarm announces an alarm in the area that it covers
   - *Remote monitoring*—When activated, a local alarm will be monitored remotely by building or campus personnel through a communications network
   - *Supervised station*—When activated, a local alarm will be monitored by an offsite company that provides recording, supervision and management of the fire alarm system

Communications and IP

The fire alarm system must call out in case of alarm. The NFPA code requires that a fire alarm's digital alarm communicator/transmitters (DACTs) be connected to two independent means of communication to a supervising station. Traditional fire alarm communications have used two dedicated phone lines to call out to a supervising station or a primary telephone line with the second method of communication being a cellular telephone, active multiplex, derived local channel, one-way private microwave or one-way or two-way radio frequency link.

In 2002 the NFPA 72 code adopted standards for IP technologies and later stipulated requirements for use of other communications technologies including a packet-switched data network (PSDN) such as IP and Ethernet. Internet connections are now being used for one and possibly both communications paths depending on the local AHJ. The Internet connection is faster, more secure and costs less.

The Internet connection to fire alarm systems uses a dual-destination IP receiver address, which provides for redundancy, can be encrypted, and can be tested on a regular and frequent basis. Some central station receivers can support up to three destination IP receiver addresses for extra equipment redundancies providing more configuration options.

In addition to supplying another type of communications to the supervising station, the two-way communications of an IP network are used by technicians and managers to upload and download information to the control panel, such as downloading alarm panel data, uploading program updates or performing system tests remotely. Fire alarm systems are starting to deploy some use of IP communications protocols in other areas such as between fire panels.

There is still substantial reliance on communications protocols developed for building automation systems, such as BACnet and LonTalk; however, these protocols can be easily routed to an IP system and embellishments of some protocols, such as BACnet/IP, are evolving as the dominance of the IP protocol is becoming more widely recognized.

Mass Notification Systems

Mass notification systems (MNS) are used to provide real-time descriptive information and directions to people during fire and nonfire emergencies. For example, many fire alarm systems have a paging system to notify building occupants of a fire situation en masse. Mass notification systems cover not only buildings, but campuses, cities, regions, and the globe; thus, MNS types are for buildings, wide-area notification, and distributed notification. MNS are used for fire situations, public alerts, emergency situation conditions and warnings such as for severe weather.

MNS use a range of technologies to provide information and instructions to people including speakers, electronic digital-message displays, computer interfaces (desktop alerts), reverse 911, SMS text messaging, commercial radio broadcast, cable TV, PDAs, cell phones, digital signage, strobe lights, and paging.

Figure 9.3 LED signage for emergency notification.

A typical MNS includes a central server or control unit, a system administrative terminal or operating consoles, and a network of notification devices such as speakers and digital signage. The system may be connected to the Internet or a cellular service to broadcast notifications via those media. Outdoor systems may include outdoor speakers and sirens providing voice signals or alarm tones such as Federal Emergency Management Agency (FEMA) weather warning tones. MNS have now been mandated in all U.S. Department of Defense facilities (Fig. 9.3).

The MNS server or control unit, along with the administrative terminal, is used to monitor and control notifications, that is, it generates and sends live or prerecorded messages. The control unit is typically integrated with the fire alarm control panel (FACP).

NFPA 72 has specific requirements for MNS and fire alarm systems. As of this time, details of the design, installation, and testing requirements for MNS are being developed for incorporation into a new chapter called Emergency Communications Systems (ECS). The new requirements include risk analysis of the probability and potential severity of events requiring emergency response. There is also a requirement for two-way telephone radio communications service addressing the emergency responders on the ground communicating with those in the building.

IP PAGING SYSTEMS

Legacy paging systems were typically a separate piece of equipment connected to a PBX. Speakers were daisy chained in a paging "zone" and connected back to the paging amplifier. Access to the paging systems was through a microphone or a special call group in the telephone system.

Current systems are built on VoIP-type technology and IP end devices including IP speakers. The network connectivity allows authorized users to send and broadcast audio simultaneously to speakers and IP telephones. Some systems have the capability to concurrently send a multicast audio stream and text messages that can be delivered to not only paging speakers and IP telephones but also PCs and non-VoIP telephones.

Authorized users can create paging groups which are similar to the legacy paging zones, but with much more functionality, thus allowing users to select particular end devices for the group. Users can control broadcasts from a PC or a web browser on an IP phone and send a live, recorded, or scheduled broadcast to one or more paging groups.

The paging system uses IP-addressable speakers with the average speaker utilizing power over Ethernet (POE) thus eliminating the need for local power. Paging horns, which require about 20 watts, will need POE Plus. All speakers can be centrally controlled and managed via the network.

Some manufacturers of IP paging systems have integrated their products into contact closure devices, generally associated with an access control system to monitor which doors are open or closed. The contact closure at a door is integrated with the paging system so that an open contact or open door can trigger a page.

Chapter 10

Voice Networks and Distributed Antenna Systems

Overview

Wired telephone service in buildings or organizations has undergone a technological revolution over the last several decades. Beginning in the late 1970s,
large telephone systems used a technology called "time division multiplexing"
(TDM).

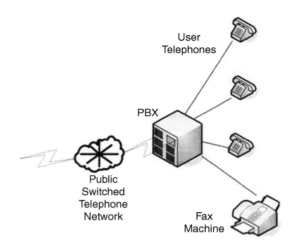

Figure 10.1 TDM PBX telephone system.

During those years, the market was dominated by a few large companies with systems open enough to connect to the outside world but that had proprietary operating systems and core hardware. When deployed for larger installations, those systems are referred to as private branch exchanges (PBXs) and are essentially privately owned hardware that could connect and exchange calls with the outside world (Fig. 10.1).

Voice over IP

The market for telephone systems significantly changed in the late 1990s with the introduction of voice over IP (VoIP) technology. VoIP essentially utilizes a data network based on the IP to transmit voice. It encodes analog voice into digital data packets at one end and decodes the digital data packets into analog voice at the other end. As the marketplace changed hybrid systems based on both TDM and VoIP technology became available. The hybrid systems are transitional systems allowing users to gradually evolve to VoIP technology.

VoIP has many advantages over the older TDM technology. A main selling point of VoIP is that, as voice essentially becomes an application on a data network organizations need only deploy a single network rather than separate networks for data and voice. This results in savings, primarily related to the consolidation of resources for management, administration, and maintenance of two networks.

VoIP offers more functions and features through the true integration of voice and data networks. Conventional TDM telephone systems had attempted to integrate telephone systems and data networks but the functionality of VoIP in this area far exceeds anything previously developed. For example, when a caller leaves a voice-mail message for a VoIP user the voice mail can be automatically digitized as an audio file, attached to an email and sent to the user. A user can listen to and/or file the voice-mail message on their personal computer or other receiving device.

Another example is a customer using a company's web page to obtain customer service; if customers decide that they need to talk with a customer service agent, they can do so by just clicking on an icon and instantly engaging in a voice-to-voice customer service chat.

Many VoIP telephones are capable of being easily relocated. Relocation only requires network or Internet access. Phones connected to an organization's VoIP system can move from an office to a home or to a hotel room through the use of a virtual private network (VPN) established over an Internet connection. The VoIP system can track where the VoIP telephone is and forward calls, providing the appropriate calling privileges and features with the telephone always operating in the same manner as if it were in the office.

Issues related to VoIP are primarily concerned with the ability to deliver voice transmission in real time. Conventional TDM systems essentially establish a circuit between the caller and the party being called with the circuit reserved solely for that particular voice call. The VoIP world is different.

VoIP encodes voice into data packets. Data packets are then transmitted over the network in which the data packets compete for network resources and bandwidth. The competition of the data packets for network resources can produce delays in the transmission and even loss of or errors in the data packets. The result of such effects is degradation of voice transmission. To compensate, the data network must provide a certain "quality of service" (QoS) for the voice transmission. This is accomplished by the data network determining which packets are for voice transmission and prioritizing them, then dedicating network resources through hardware or software for the voice packets.

VoIP technology essentially takes voice (an analog signal) and encodes it to a digital signal to transmit over the network. You hear the person you are conversing with through the reverse process, which is the digital signal decoded into analog. The signal is encoded and decoded and compressed through software or hardware called a "codec," and then packetized in IP or Ethernet packets.

Encoding analog to digital signals occurs at different rates, trading off the quality of the voice signal with the network bandwidth to carry the signal. The International Telecommunication Union's G.711 is the most widely used standard, compressing voice at 64 Kbps.

There are also several communications protocols used for VoIP technology that address signaling and voice transmission. These addressing signaling protocols include the Session Initiation Protocol (SIP), the Media Gateway Control Protocol (MGCP) and H.323. The SIP initiates, modifies, and terminates sessions with call participants and is integral to providing call services such as call forwarding and caller identification.

The MGCP is primarily used for controlling telephony gateways—devices that convert analog signals to digital signals such as analog telephone service from a telephone service provider connected to a VoIP system. The H.323 standard is an umbrella of standards focusing on audio, video, and data communications across IP-based networks that do not provide a guaranteed QoS.

For voice transmission the standard protocol is the Real-time Transport Protocol (RTP). This protocol provides end-to-end network transport functions, such as control and identification and is supplemented by a control protocol (RTCP) to allow monitoring of data or digital voice delivery.

Another VoIP issue has been providing power to VoIP telephones. In a conventional TDM system power is centralized and feeds each of the instruments attached to the system. In early deployments of VoIP, VoIP telephones had to be powered locally, at the location of the telephone instrument. This meant that not only did the telephone need a network connection but it also required a power outlet, thus adding cost to VoIP deployments and making the moving, adding, and changing of instruments cumbersome.

The market response was to develop a method to provide power to the telephone instruments through the network connection. It resulted in the IEEE 802.3af standard that set guidelines for providing power over a network cable (commonly referred to as POE) and discussed previously.

Telephone users can use analog or digital telephones. On conventional TDM telephone systems standard analog telephones or proprietary digital instruments would be used. On VoIP systems, analog telephones can be used with an adapter at individual instruments or a network gateway that converts multiple analog telephones to the digital network. Other analog devices, such as fax machines, are serviced in the same manner as analog telephones.

Digital telephones for a VoIP system can connect directly to a LAN network switch, much like a desktop PC would be connected. Many digital telephones come equipped with a built-in "mini-data switch," allowing one network connection to be connected to the instrument which serves both the telephone instrument and a desktop PC connected to the mini-switch. Softphones are another option for VoIP systems.

Softphones are software applications for a PC that essentially turn it into a telephone. Users utilize a headset or a USB-connected telephone to complement the software application, resulting in a full-feature telephone set. Software on PCs allows for onscreen dialing and access to a user's contact lists.

Unlike conventional TDM telephones VoIP telephone instruments are really network appliances. A VoIP telephone with an LCD display allows users to browse the Internet or access an organization's network, including applications and databases. With the use of XML, the VoIP system can provide the user with access to many applications such as news, stock market reports, and weather and telephone listings. The applications on the VoIP telephone can be tailored to building use. For example, in a hotel the VoIP telephone may allow a guest to order room service or check out or it may provide for hotel housekeeping or engineering staff to enter or access data.

More critical to a smart building is the VoIP's capability to integrate into building systems such as HVAC, lighting control, access control, and video surveillance cameras. This allows users control of their environment by setting temperatures, lighting schedules, lighting levels, opening and closing window blinds.

It is clear that VoIP will be the dominant future technology used in telephone systems. With telephone systems mimicking and riding on data networks and utilizing similar standard cabling infrastructure and dominant network protocols, they can easily be integrated and converged into other smart building systems.

Distributed Antenna System

Understanding wireless communication systems in buildings can be confusing as there is a range of different systems and applications and radio frequencies for the systems. In addition, an established business model for systems deployment or a clear contractual model for providing certain wireless services does not exist. Buildings may need cellular coverage, a Wi-Fi system, extended wireless public safety communications, radio frequency identification (RFID) systems, paging and other wireless devices. For building owners, in-building wireless systems can have several benefits: greater tenant and user satisfaction, fewer facility operational problems, increased public safety capabilities and improved tenant amenities.

Among the wireless services cellular service has become the most important in buildings because of high penetration of cell phone usage in the population, reliance by many on the cell phone as a primary phone, and the added importance of receiving and transmitting email and text. Some cell carriers report that around 60% of the telephone calls have some "indoor component."

The result is that the user's tolerance of dropped calls inside a building is very low.

Cell services to most buildings are provided by the closest outdoor cell base station or tower. Several elements affect in-building cellular service including building structure and materials, distance to the nearest cell tower and location of the caller in the building. Users at the upper floors of a high-rise building can have problems as their handsets continuously hunt between multiple cell towers. Users in underground floors may have no coverage at all. Technically, the solution to dropped cell calls is to essentially "extend" a carrier's cell service antenna into the building. This can be done in different ways.

One method is to mount a small directional antenna on the building pointed toward a particular service provider's cellular antenna; this is typically referred to as the "donor" cell site. The building antenna provides direct two-way communications with the cell tower. An alternative, albeit probably more expensive option, is to have a "land line" connect the building and the cell tower such as a T-1 line or metro Ethernet connection between the two locations.

From the antenna on the building coaxial cable is run to amplifiers that are usually located in the main telecommunications equipment room. Coaxial cable is commonly used to transmit RF and is well suited for transmitting RF between the outdoor building antenna and the building distribution amplifier. Because coaxial cable may run from outdoors to indoors, the installation requires proper grounding and lightning arrestors.

From the amplifier in the telecom room a distributed antenna system (DAS) is installed throughout the building. The in-building distribution system comprises a series of strategically located indoor "omni" antennas connected through coaxial cable using signal splitters and couplers similar to a cable television distribution system.

Additional equipment, such as inline amplifiers and boosters, can be used in the distribution system to address the loss of signal strength through the coaxial cable. In a campus environment, the amplified signal from the antenna can be transmitted over the campus backbone network to other buildings.

Signal strength and antenna coverage of the distribution system will be determined by building structure, power of the indoor antennas, frequency band of the cellular carrier and loss of signal through the coaxial. All of these factors can be accounted for in system design and engineering (Fig. 10.2).

The early DASs were based on a single cell carrier and the exclusive use of coaxial cable. More recent systems handle multiple carriers and distribute the signal using managed network hubs that provide the same signal strength at every antenna.

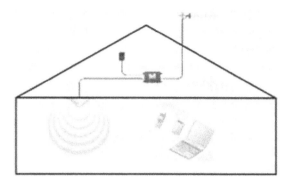

Figure 10.2 Simple distribution antenna system.

This technical scenario works not only for cellular service but also other wireless services. For example, the RFs used for public safety can be "extended" into a building's DAS allowing public emergency responders to readily communicate. The DAS-type infrastructure for cellular services can also be leveraged to provide in-building Wi-Fi. Some commercial real estate owners are finding that complete in-building Wi-Fi is less expensive and presents fewer issues than dealing with tenant-provided Wi-Fi systems that interfere with each other.

Pico Cells

Pico cell systems are small radios, similar to the outdoor cellular base stations, but limited in capacity and power. They are manufactured by the major cell equipment companies and are best suited for apartment complexes and small office buildings. They are typically connected back to a cell base station using an IP network; the base station controls the pico cells, thus managing the communication traffic to and from the small cells in the buildings.

The pico cells can compete with the outdoor macro cell for coverage. That is, a cellular handset could possibly hunt from the small pico cells to the macro cells for service. To prevent this, the pico cells have to provide a strong and dominant signal. The strength of their signal affects the area that will be covered and the system's caller capacity.

Multiple pico cells used in a building can provide better coverage but because each cell uses the same frequency more interference or noise is created. The result is decreased capacity of the system to carry higher data rates. Pico cells combined with a DAS can provide better overall system performance.

Business Issues

Like most decisions about building technology systems, in-building wireless is less about the technology and more about the business case. The business issues surrounding in-building wireless involve several parties: the building owner, cell phone carriers, and third-party companies that install and facilitate the improved cell services.

Carriers are interested in in-building systems because they can increase billable cell minutes, reduce subscriber dissatisfaction due to poor coverage and help retain customers. A carrier will pay for deploying an in-building system depending on the estimated increase in business. The carrier will consider the installation cost, customer base, the projected increase in sales and the time for return on investment.

In commercial real estate, large tenants may demand cell coverage in their space or a property manager may install a DAS to improve their competitiveness with other properties or to support their leasing rates.

Third-party companies providing in-building cellular service will partner with or manage the relationship with the cell carrier.

The key business issues are who "owns" the customer, who is making the capital investment in an in-building system, and how revenues from the system are distributed. Cellular carriers want to control their networks and their customers. Some of that control may be lost depending on how an in-building system is deployed. Other business issues that arise include revenue sharing between carriers and third parties and who's responsible for the customers' primary contact.

Emerging Technology Trends

The convergence of cell service and Wi-Fi will change business dynamics. Handsets capable of cell and Wi-Fi are available, allowing users to get voice, data and video service from a single handset through two different networks. Users can now choose whether to make a call via cell services or a VoIP service, such as Skype, if within Wi-Fi coverage.

The most logical approach for most building owners would be to deploy a single in-building wireless system that could at minimum handle cellular, Wi-Fi, and public safety. Such a system would improve tenant amenities and public safety capabilities while possibly avoiding operational issues resulting in added value for the building.

Chapter 11

Data Networks

Overview

Data networks are particularly important for smart buildings. The basic infrastructure of data networks (standard cable infrastructure, using the IP network protocol, interoperable databases) is open and standardized and is proliferating and being adopted by other building systems. The basic data networking technology infrastructure comprises the technical core of a smart building.

Doi:10.1016/B978-1-85617-653-8.00011-9

Data networks are used to share resources and exchange information between network users and other networks. Decades ago data networks consisted of mainframes, minicomputers and proprietary networking infrastructure and communications protocols. Today most networks consist primarily of switches, servers, industry-standard operating systems, network and client software applications, peripheral devices and user devices.

Users typically use desktop or notebook forms of personal computers to connect to a data network, although some of the basic PC applications are now accessible with smart cell phones. The network connectivity of the PC user to the network is either structured cable or a wireless access point. Remote connectivity to the network is typically provided through a telecommunications service provider via a cable modem, DSL, T-1 lines or higher-capacity telecommunications services.

Networks

A local area network (LAN) is typically deployed for a building or a cluster of network users that have similar needs such as an administration department, facility management department or computer laboratory (Fig. 11.1).

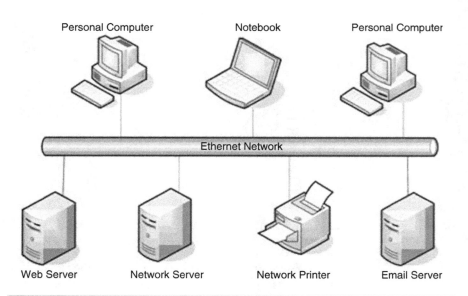

Figure 11.1 Local area network.

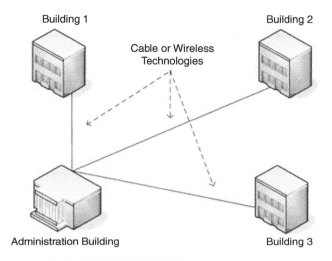

Figure 11.2 Campus network.

A campus network serving a university, corporation, or government connects the LANs within each of the buildings to provide a "seamless" campus network. Campus networks have connectivity between buildings provided via cable, although alternatives include the use of wireless and telecommunications circuits (Fig. 11.2).

Data networks can scale to metropolitan areas and wide area networks covering countries, continents, and the world. Connectivity between sites can be provided via point-to-point terrestrial (land lines) or nonterrestrial (such as satellite or wireless) telecommunications circuits, the Internet, or some mix of connectivity.

A network consists of three general device types described in the following sections.

PERSONAL COMPUTERS OR OTHER USER DEVICES

A user's personal computer has its own processing, memory, storage, operating system, and software applications and can operate independently of other systems or networks. However connecting the PC to a network allows it access to other resources including printers, the Internet and centralized databases or software.

Typical desktop computers and laptops access a network through an Ethernet cable, wireless device, or a telecommunications circuit. The PCs will have network adapter or a network interface card that provides the physical access to a network or the Internet.

The network adapter or interface has a unique identifier called the media access control (MAC) address assigned by the manufacturer. There are variations of user devices and personal computers typically used on a network, such as "thin clients" (essentially a PC without any local storage) and network "appliances" (devices geared to a specific application, such as Internet access and email).

NETWORK SWITCHES

Network switches are the "traffic cops" of the network, connecting Ethernet network segments. The switch allows each device that is connected to have dedicated bandwidth to the switch without any potential for "collisions" of data packets. When devices on the network (users and servers) transmit information over the network, the switch determines the origin of the message or data packet, its destination and the best path or route to get the message to its destination.

Most switches will handle multiple variations of Ethernet, with transmission speeds of 10/100/1000 Mbps. Large, more sophisticated switches may have 10 Gbps ports. Network ports for other types of networks such as Fibre Channel or asynchronous transfer mode (ATM) may also be available. Typical switches will provide for ports in multiples of eight: 8-ports, 16-ports, 24-ports, 48-ports, and so on.

Network switches may operate at one or multiple open system interconnection (OSI) layers: physical layer, data link layer, network layer or transport layer.

Layer 1

Network hubs deal strictly with Layer 1, the physical layer. Hubs simply repeat or broadcast out traffic to all other ports. Hubs do not manage any traffic and therefore "packet collisions" can result and affect throughput. In some cases it is advantageous to broadcast traffic to other ports, but more sophisticated switches can do something similar with port mirroring.

Layer 2

The network equipment that manages Layer 2 connectivity is referred to as bridges. Bridges provide only one path between any two points using a "spanning tree protocol" to learn the network topology and then find the best path. The bridge then forwards the frame or packet using one of the following methods:

1. Store and forward, which buffers and checks the frame prior to forwarding

2. Cut-through, which only reads the hardware or MAC address prior to forwarding

3. Fragment free, which is a combination of store-and-forward and cut-through that reads the physical address and performs some error checking

4. Adaptive switching, which switches among the first three methods as appropriate

Layer 3

The network layer of the seven-layer OSI model is responsible for end-to-end packet delivery. Whereas the previous layers may deal with activities with a network, the network layer involves communications via one or more networks. Forwarding data packets to different types of network connections is usually the function of a router, although a Layer 3 network switch can perform all the functions of a router.

Layer 4

While Layer 4 switches are not well defined they do have some traits common to the other layers. One is the capability to perform network address translation (NAT). Typically NAT is used to allow multiple users or devices to access the Internet using one public IP address. Years ago this type of translation was a tool to deal with the unavailability of IPv4 addresses. More recently it is used to hide an internal network structure by making it appear that all network traffic is originating from the Layer 4 switch rather than the devices on the internal network.

Because NAT involves translating IP addresses more processing is required and the model of the end-to-end connectivity across the Internet is somewhat blemished. Layer 4 switches may also do load balancing. With this feature the switch resource utilization and traffic throughput is optimized. An example is a server farm where the network switch may direct traffic across several servers to balance the load. The switch uses policies or filters to identify and manage application specific traffic.

NETWORK SERVERS

Servers are connected to LANs and provide a variety of resources to both network users and network administrators. A common application provides network users with connectivity to the Internet and shared Internet firewalls. Servers have many other uses including hosting web pages and email applications, centralized databases or software applications, printing applications and network administration capabilities. Thus the categories for servers are database servers, applications servers, communications servers, file servers, print servers, proxy servers, and web servers.

From a hardware perspective servers must operate under the heavy demand of a network environment. Their configurations can vary from a desktop PC to a mainframe computer. Most servers are built for a production environment with fast CPUs, multiple processors, hardware redundancy in disks and power supplies, large storage capacity and specialized operating systems and software applications. A server may be installed in an equipment rack or for high density may be a "blade" in a server chassis mounted in an equipment rack.

IP Addressing

Each physical connection of a device to an IP network has an IP address. The IP address is a numerical identification and a logical address of the device connected to the network. It is used to specify the routing source and destination of the transmission over the network. Originally an IP address was defined as a 32-bit (or binary numbers, 1s and 0s) number, a scheme referred to as Internet Protocol Version 4 (IPv4). Because of the growth of the Internet and the limits of IPv4, a new scheme was created to handle many more IP addresses. This is called Internet Protocol Version 6 (IPv6) and uses 128 bits for the IP address.

There is also a set of IP addresses set aside for use in private or closed networks. Anyone can use these addresses in their private networks, probably supplemented with a translator to go from the private network to a public network such as the Internet. An example of an IP address is "162.146.93.14." Behind the scenes that example IP address would be represented to a network in a 32-bit binary format: "10100010.10010010.01011101.00001110."

There are five "primary" classes of IP addresses. The first three, Classes A, B, and C can progressively handle more networks. Classes D and E are for multicasting and research.

All Class A addresses and almost all Class B addresses have been allocated. One way for network administrators to handle the limits of the IP addresses allocated is to subnet or create subnet IP addresses within the IP addresses allocated.

Integrating technology systems in buildings is analogous to a data network in that there is a "core" network in which there is commonality in the physical connectivity and the way that the systems communicate. Outside of the core network are various devices and controllers made by different manufacturers with different capabilities and functions. This type of standardization has significant functional and cost implications which comprise the foundation of the deployment and benefits of smart building technologies.

Chapter 12

Facility Management Systems

Overview

A facility management system (FMS) is an overarching system of a smart building that brings together some of the operational management functions of the facility and the building technology systems. The FMS is typically a server-based configuration coupled with operator workstations, which may be supplemented with wireless devices.

Access to many FMS can also be achieved through Internet access. FMS environments typically operate on a standard Ethernet IP network using a structured cable infrastructure and industry-standard operating systems and databases.

The definition of an FMS can be confusing, especially in comparison to a building management system (BMS). An FMS focuses on the business processes of facility management. It is a tool that assists in managing service orders, inventory, procurement and assets. These systems are generally vended by companies with specific applications for facility management operations or by companies that are involved in broader business process products such as human resources, finance, purchasing, and so forth (Fig. 12.1).

A BMS is focused on the operational functions of the smart building systems, primarily life safety and building automation systems. The BMS is typically vended by the manufacturer of the building automation and life safety systems. These systems may integrate systems and controls, provide data on specific devices or equipment, provide system alarms and allow the operators to establish set points and system schedules, among other functions.

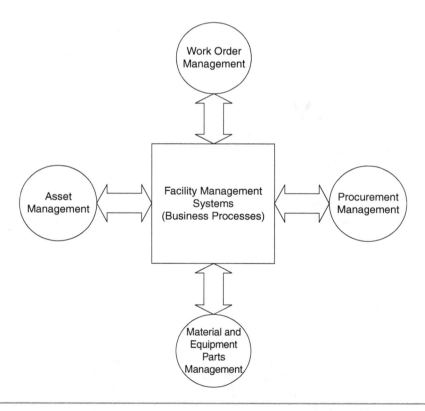

Figure 12.1 Facility management system.

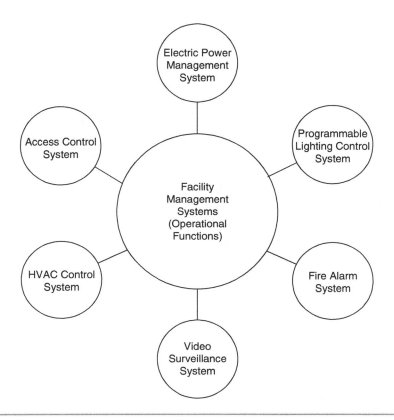

Figure 12.2 Building management system.

There may be a bit of overlap between the FMS and BMS. For example, both may provide for some type of energy consumption reporting. In many cases however facilities need both an FMS and a BMS (Fig. 12.2).

Facility Management Software

The FMS consists of a series of software modules that allow an owner or facility manager to select modules based on operational and facility management needs. Modules that may be included are described in following sections.

WORK ORDER MANAGEMENT

A work order module tracks and monitors all service requests. A work order management application:

- Tracks people assigned to specific actions

- Tracks current status of actions
- Keeps records of action updates
- Tracks when actions are completed
- Archives work for future reference

Through the use of wireless devices a system can notify technicians and other mobile employees with service requests or other work order data on any text-messaging or email device such as cell phones, tablet personal computers, or pagers. Systems may allow for tenants or occupants to send service requests over a network or the Internet and be automatically notified on the status of the service request.

A work order system tracks labor, materials, and travel expenses for work orders. It can also attach documents to work orders, project labor demands for any future time period based on the work orders and monitor work-performance indicators in addition to cost or time.

Asset Management

An asset management module manages all necessary equipment data including names and models, serial numbers, location, vendor, internal cost center, warranties, performance and documentation. The company or organization may have an asset management program that is part of their enterprise resource planning (ERP) which is a business-level software platform, and in this case the data on the assets of building systems and operation may reside in the ERP.

Material and Equipment Parts Management

A material management system tracks material and equipment parts that are moved into and out of inventory, and that currently reside in inventory. It ensures that the right level of inventory is maintained. The material management module is typically tied to and integrated with the work order system and purchasing system.

When a work order is generated the FMS may check the inventory for the parts needed for the repair or to complete the work order. The inventory system may have set thresholds for various parts and materials and automatically triggers the reorder and restocking of parts and materials through the purchasing system.

PROCUREMENT MANAGEMENT

Procurement management automates and streamlines the procurement process for vendor services and equipment. Contractors are preapproved for specific types of services and equipment. Then for example, the work order management system can automatically assign a work request to a preapproved external contractor triggering and tracking the procurement process needed.

Procurement management has other benefits. Some systems can track contracts, key contract dates and contractor performance. Systems can also automatically requisition materials and equipment based on predetermined maintenance schedules.

Building Management Systems

A BMS monitors, supervises, controls and reports on smart building technology systems. These systems may include access control, video surveillance, fire alarms, HVAC control, programmable lighting and electric power management. Its basic functions include:

- Providing information on supervised building functions including, but not limited to, current status, archived historical information, summaries, analysis, displays, and reports on control and management functions

- Detecting, annunciating, and managing alarm and other conditions

- Diagnostic monitoring and reporting of system functions, nodes, devices, and communication networks

- Interfacing between individual smart building applications

These systems typically display the following responses on an operator's workstation:

- Alarm summary
- Event summary
- Trend set displays
- Group control and group trend displays
- Communications status
- System status
- Configuration displays
- Communication links status
- System parameters configuration

- Time schedule assignment
- Holiday assignment
- History assignment
- Events archive and retrieval
- Time period summary and configuration
- Point details for every configured point

Energy Management System

An energy management system (EMS) generates information on energy usage and related costs for the purpose of reducing costs while still maintaining a comfortable and safe environment for building occupants. As part of a smart building the EMS brings together and addresses the main electric and energy systems, namely HVAC, lighting control and power management.

Electrical utilities base their charges on several factors, but the most important are power consumption and demand. Consumption is simply the total amount of electricity used in a billing period. Demand is typically cost per kilowatt depending on time or season (Fig. 12.3). So reducing consumption and managing demand are basic strategies of an energy management program. An energy management program can coordinate the HVAC and lighting control systems together with an equipment maintenance program to achieve optimum energy usage.

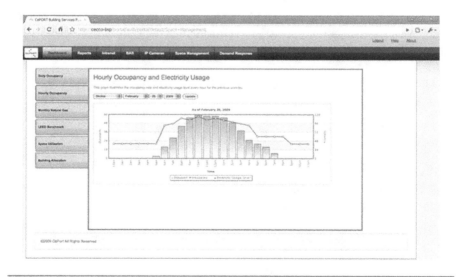

Figure 12.3 Example of energy management information.

The EMS may be a software module in an FMS or BMS or a separate stand-alone application. An EMS usually provides a group of applications to optimize a facility's energy usage and costs. This may include the following programs:

- Tracking of utility bills to monitor usage and costs as well as comparing projected, budgeted and actual usage and costs
- Comparing the energy usage to other similar buildings, an exercise known as "benchmarking"
- Calculating the effect of thermal comfort or lighting alternatives. That is, the cost for different lighting control strategies, proper ventilation rates, and related elements

Because energy can be saved by ensuring that equipment is operating in the most cost-effective manner an EMS may have a feature to ensure proper maintenance of mechanical and electrical equipment. Maintenance can be scheduled at regular intervals based on historical data or manufacturer recommendations, or equipment can be monitored to determine abnormal operating conditions.

Monitoring requires setting acceptable operating ranges for equipment and then setting alarms for when the equipment operates out of range. Equipment indicators that may signal possible malfunctions include the following:

- Temperature
- Vibration
- Pressure
- Air quality
- Humidity
- Energy consumption
- Gases

Computerized Maintenance Management Systems

Computerized maintenance management systems (CMMS) are comprehensive software applications for a variety of equipment and materials. Like EMSs, a CMMS implementation can be a standalone application or integrated with a software module of a larger FMS. The CMMS is probably the most essential and expensive part of an FMS.

The CMMS integrates or communicates with the FMS to retrieve data from field devices in determining when equipment or materials require maintenance.

The CMMS triggers an automatically generated work order request for a piece of equipment when certain maintenance conditions are met. Configurable maintenance event types may include duty cycles, run hours and high data values.

A CMMS can include a fleet management module that automatically creates work orders for preventative maintenance of vehicles; it can establish a vehicle replacement schedule based on vehicle cost, as well as track registration, license renewal, and warranty work.

A CMMS software package can provide information for other management areas and tasks. These include property and lease management, budgeting, space and moving management, room reservations, furniture and equipment management, emergency preparedness, environmental quality, and maintenance the work order requests. A feature of certain recent CMMS programs is the capability to link data to spaces within a building. This is typically done by a CMMS that can handle Computer-Aided-Design (CAD) files and present information in an intuitive visual manner.

Many CMMS programs can be linked to a web browser, allowing facility managers to access specific functions of the CMMS remotely with role-based security features that prevent unauthorized access to data. Off site field personnel can view work order requests from remote locations and determine which tools and materials are needed (Fig. 12.4).

Facility Manager

The position of facility manager is a long-standing profession known by many titles. While only recently professionally accredited, facility managers have a long list of increasingly important responsibilities that continually grow as technology becomes more prevalent in the workplace and the complexity of buildings increases.

The facility manager must keep pace with technology while maintaining a seamless and invisible integration of these new technologies. Facility managers must have a background in construction practices, local regulations, purchasing, project management, account management, human resources, building systems and operations.

The duties of a facility manager will depend on the size and type of facilities being managed, ranging from mechanical equipment maintenance and repair scheduling, to providing interoffice relocation support to employees. Organizations such as the International Facilities Management Association (IFMA) and Building Owners and Management Association (BOMA) provide continuing education and assist facility managers in staying abreast of information pertinent to the profession.

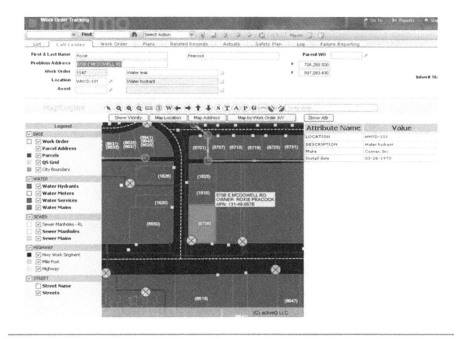

Figure 12.4   Web access to work order information.

One of the main roles of facility managers is to decide what level of system integration is cost effective for their facility. While building technology systems and the level of system integration may occur in the design phase of a new building and many decisions will be left to the architect and engineers, it is the facility manager who eventually will monitor and manage the building's performance. The facility manager must be actively involved in building design and the systems integration effort for it to succeed.

Design of the building and its systems should begin with answering questions regarding how the building should be operated and what services need to be provided. The facility manager in this regard may take on the following tasks:

- Identify the need for systems integration and the levels of integration needed
- Evaluate costs and benefits of various approaches
- Present options to decision makers in a way they can understand
- Remain part of the integration process during installation
- Be involved in commissioning and maintenance of the new systems

Chapter 13

Design, Construction, and Renovations

Overview

Decades ago, when centralized HVAC systems were invented and manufactured, their attendant needs for space, piping, ducts, and so on had a major impact on the fundamentals of building design. Technology, telecommunications, control systems, audio/visual, and security systems are having a similar impact on the design process today.

Doi:10.1016/B978-1-85617-653-8.00013-2

Like other building design disciplines, technology comprises many systems and components that need to be tightly coordinated with other design professionals on the team and integrated into their designs. Technology design in buildings as evidenced by changes in the MasterFormat is now formally incorporated into the design and construction process.

The Design and Construction Process

The process of planning, designing, constructing, and managing a building is complex. Most buildings are unique structures designed for certain functions, budgets, physical conditions, delivery methods, owner requirements, regulations, schedule constraints, and other influences. The number of people or organizations participating in the design and construction of a building is substantial.

It includes the owner's team, which may consist of the owner, a facility manager, and the tenants or those occupying the building; the design team, which may be comprised of architects, engineers and consultants; the construction team made up of a general contractor, subcontractors, and a supplier's team that may include distributors, manufacturers, and product representatives.

With such a variety of potential project influences and restrictions and a multitude of participants there has to be a methodology as to how the needs of the owner can be precisely communicated to the designers as well as how the designers will accurately communicate their design to those constructing the facility. This process consists of the following major phases: project conception, project delivery, design, bidding/negotiating/purchasing, and construction. Each phase is briefly described in the following sections.

PROJECT CONCEPTION

This is the programming phase in which the owner's needs and requirements are identified. A building site is selected and a preliminary budget and schedule are set. More importantly, a facility program, based on the owner's values or goals is prepared. The facility program is the foundation for many of the design decisions. Except in cases of unusually astute and informed owners technology and integrated systems are rarely a substantial part of the facility program.

Participants in this stage typically include the architect, engineers, facility programmers, and representatives from large owner/user groups. Oftentimes the foundations for building design are established without thorough discussion of building technology systems.

PROJECT DELIVERY

This is the "relationship" phase where the owner decides on contractual relationships among all participants to get the building from concept to completion. Common delivery methods include design-bid-build (the traditional method), design-negotiate-build (negotiating with one contractor), design-build (a single entity performs both design and construction), construction management (a third party augments the owner's role), and owner-build (everyone, including design team, contractors and subcontractors contract directly with the owner).

Although those involved in building technology systems do not participate in selecting the delivery method for the building it is important to understand how the building is being "delivered" as this will affect the relationship one has to the design team, the owner, and the construction company (Fig. 13.1).

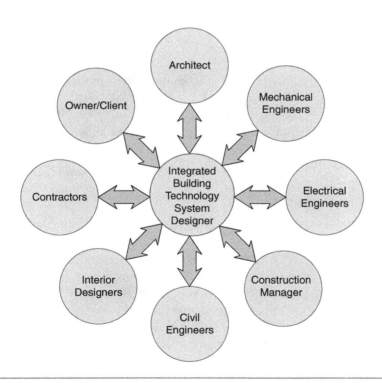

Figure 13.1 Project team.

DESIGN

When the owner's conceptual design and programming are complete and the contractual relationships among all parties established the detailed design of the building begins. The design will proceed through a two- or three-stage process. The three-stage process consists of schematic design, design development and construction documents. The two-stage process includes preliminary design and final design, leaving construction documents as a separate step. It is typically during this phase that a designer of building technology systems interacts and collaborates with many of the architects and other technical design professionals. The following sections describe some examples.

Architects

Typically, architects lead the design team. Many of the interactions that designers of building technology systems have with architects involve spaces for equipment rooms, and end-user devices and related cabling that may need to be coordinated with finishes, millwork, furniture and so on.

Civil Engineers

The civil engineer is interested in telecommunication service entrances into a building and coordination of those pathways with other utilities. Considerations for these pathways may include multiple telecommunications carriers, multiple entrances, connectivity to other buildings, among others. It may also be necessary to consider items such as pads for satellite dishes or antennas in site planning.

Mechanical Engineers

These engineers will be designing the HVAC system while the building technology systems designer is concerned with the automation system that monitors and manages the HVAC system. Thus, close coordination is required. In addition, the mechanical engineer's design must consider the heat loads and cooling needs of the technology equipment. These needs are most acute in data or network centers and telecommunications or equipment rooms where technology equipment is dense.

In many jurisdictions the mechanical engineer will supervise a plumbing engineer. The plumbing engineer may be involved in a water-based fire suppression system, irrigation system, domestic/potable water system and plumbing fixtures, such systems that are integral to life safety and resource management.

Electrical Engineers

Electrical engineers will design several systems requiring extensive coordination with the building technology system designer. These may include lighting control, fire alarm, and power management. In addition, coordination is needed for power and grounding for equipment rooms, lighting, audio/visual systems, location of power outlets and networking outlets, and so on. Typically, the electrical contractor installs cable trays and conduit so that all cable pathways for technology systems must be coordinated with the electrical engineer as well.

Owner Representatives

These representatives, including the information technology director, facility manager and the security manager, are involved in the building process because they will eventually be managing and operating the systems. The initial configuration of building technology systems, training in system management and operation, system documentation, commissioning and long-term support, and maintenance and warranty of the systems are all part of the design process that owner representatives will be involved in. In addition, an owner may have legacy systems requiring migration of older equipment to a new building or may have established relationships with manufacturers or contractors that need to be considered.

As the design progresses and becomes more specific, the designer has the responsibility to ensure that the design is constructible; in other words, that what is designed can actually be built or installed. The designer does this by communicating with potential contractors and researching and identifying products that will meet the design criteria.

By identifying selected products the designer sets performance standards, takes into account product history or life cycle, gets owner approval, further develops the design and refines project cost estimates. This may be a time for manufacturers or contractors of building technology systems to provide technical advice to the designer and assistance in specifications, drawings, and cost estimates.

System specifications will typically follow the MasterFormat by the Construction Specifications Institute (CSI). In this format major building subsystems have their own Division or Section. The older MasterFormat has 16 divisions addressing how buildings were built over the previous 10 to 15 years. Typically, when using the older format, technology-related items are provided for in a "Division 17." The newer MasterFormat contains specific divisions for building automation systems, communication networks and life safety systems.

BIDDING/NEGOTIATING/PURCHASING PHASE

In this phase construction documents become part of a procurement document and contractors are solicited to formally respond to the procurement. The designer of the integrated building systems will participate in the process by answering questions, providing clarification, and eventually evaluating and recommending contractors for the work.

Depending on the method used to specify the technology systems and corresponding funding sources only some of the systems' components, such as the cabling infrastructure, may be included in the construction documents as a part of the construction budget. Other components, such as the networking equipment, may not be included in the construction documents or construction budget and may be funded by the owner directly or a within a budget for furniture, fixtures, and equipment (FFE). If this is the case, the system designer will prepare specifications and bidding documents outside the construction specifications and may bid the equipment later in the process to guarantee acquisition of the most recent releases of hardware and software.

CONSTRUCTION

Building construction involves everyone: owner, architect/engineer team, consultants, product representatives, authorities having jurisdiction over the construction, among others. The integrated building technology system designer conducts contract administration, essentially ensuring that the systems installed are in accordance with the specifications, drawings and contract. This involves review of contractor submittals, responding to information requests, observation reports of the installation, review of acceptance tests, commissioning, and so on.

During the last stages of construction the technology designer and contractor may be involved with the client's information technology (IT) department to assist in migration of services and equipment to the facility. In addition, they will be involved with the client's vendors for telecommunications services, furniture, moving and so on.

A couple of tools are available to help with coordination between designers and trades representatives. One such tool is called "coordination drawings" which are typically prepared by the contractor and submitted to the designer. These drawings show coordination of all trades' installations of the diverse systems or equipment.

The other tool is called a "matrix of responsibilities" which is prepared by the designer during the design process. The matrix delineates the division of work across various specification divisions and systems. It identifies who will be responsible for furnishing the equipment, installing the equipment, cabling to the equipment, providing power to the equipment and providing "rough-in" to the equipment.

Figure 13.2 shows an example of specific section work, and a matrix of which division will be responsible for furnishing, installing, rough-in, wiring and supplying power.

It is this design and construction process that will shape the building technology systems and ultimately affect how the facility will be managed and how users of the facility will work, live and communicate. Understanding the process is important to designers and contractors of integrated systems if the benefits of the systems are to be realized.

RESPONSIBILITY MATRIX						
Section Reference	Work	Furnish	Install	Wire	Rough-In	Power
11XXX	Projection screens	11	11	11	16	16
15XXX	Variable frequency drives	15	15	16	16	16
15XXX	Smoke damper, fire–smoke dampers	15	15	17	17	17
16XXX	Electric Power Management System Equipment	16	16	16/17	16	16
16XXX	Low voltage lighting control panels and remote controls	17	16	17 low volt/16 line volt	16	16
17XXX	Intrabuilding data networking equipment	17	17	17	16	16
17XXX	Central controllers, card readers, cards, badging stations, etc.	17	17	17	16	16
8XXX	Door Hardware	8	8	17	16	16

Figure 13.2   Responsibility matrix.

Construction Documents

The design process for building technology systems is communicated to the installing contractor and others through specifications and drawings which together are referred to as the "construction documents." These documents are critical to manufacturers, contractors and designers of integrated building systems.

Construction documents for integrated building systems are prepared by a designer or engineer and are used to coordinate the design with other professionals, used by the contractor to install the systems, and eventually used or archived by the owner or the facility manager. The construction documents, comprised of plans and specifications, delineate the design detail and requirements for the installation.

Construction documents are binding legal documents. The contractor is bound by law to install the systems as per the construction documents. These documents identify the responsibilities of the contractor, the interrelationships between the contractors and other parties involved in the construction and the contractor's rights. In some cases (i.e., fire alarm systems) the construction documents are used to obtain government permits for system installation.

It is important to note that construction specifications are not the same as a manufacturer's product specifications. As an example, a manufacturer's specification for a video surveillance camera differs from the construction specification for a video surveillance system. The construction specifications will describe the work and the required results, quality, installation practices, materials, coordination, and documentation of the work.

The designer's responsibilities in preparing the specifications are clarity, succinctness (to the point), and technical soundness. If a specification uses words such as "any" or "all" or phrases such as "as required" or "as appropriate," it is not conveying specific information to the contractor.

SPECIFICATIONS

Construction documents have been standardized for ease of use to reduce errors and to facilitate coordination between parties. The written specifications contain divisions for every building discipline (civil, mechanical, electrical, etc.). They use something akin to a "Dewey decimal system" that is used in libraries with the specifications having standard titles, a master list of numbers and a standardized page format. Major divisions of the specifications are further divided into sections.

One of the most popular formats for construction specifications is the Master-Format, produced and trademarked by the CSI, and primarily used in North America. European and Asian entities have very similar standardized formats.

There has been effort to review and revise the MasterFormat about every 7 years to reflect new design and construction requirements. For example, the previous version of the MasterFormat had 16 divisions, but scant mention of technology. This spurred the use of the "Division 17" for almost anything that did not fit into the other specification divisions. Technology was one of the factors considered in the latest revision of the MasterFormat which was released in 2004.

The MasterFormat contains a "Procurement and Contracting Requirements" and a "Specifications" group. The Specification group has the following subgroups:

- General Requirements
- Facility Construction
- Facility Services
- Site and Infrastructure
- Process Equipment

The designs of integrated systems reside in the Facility Services subgroup. This subgroup has seven Divisions, three of which (25, 27, and 28) are related to building technology systems:

21—Fire Suppression

22—Plumbing

23—Heating Ventilating and Air Conditioning

25—Integrated Automation

26—Electrical

27—Communications

28—Electronic Safety and Security

Division 25—Integrated Automation

This division covers building automation systems including facility equipment, conveying equipment, fire-suppression systems, plumbing, HVAC, and electrical systems. It addresses the instrumentation and terminal devices for these systems as well as network equipment such as network devices, gateways, control and monitoring equipment, local control units, and software.

In addition, the division deals with system operation, maintenance, sequence of operation, schedules, cabling and cable pathways, commissioning as well as integration with other related Divisions such as Communications and Electronic Safety and Security systems.

Division 27—Communications

This division basically covers traditional IT items (structured cable, data and voice systems) as well as audio visual and specialty communications systems. It also deals with inside and outside cable and cable pathways, telecommunications services, and gets as granular as printers, virus protection software, disaster recovery, and even virtual reality equipment. This is also the division that covers a wide range of specialty systems such as nurse call, sound masking, RFID, intercom and paging systems, digital signage, clocks, and point-of-sale, among others.

Division 28—Electronic Safety and Security

This is essentially the life safety division where access control, video surveillance, intrusion detection, detection of radiation, fuel, refrigerant and gas, and fire alarm detection reside. Each of the sections within a division is presented or written in three parts:

General—This part lays out the administrative and procedural requirements for the contractor on the job.

Products—This part lists the equipment, materials and products required for the system and job.

Execution—This part describes how the products and equipment are to be installed, post-installation requirements, documentation, and so on.

DRAWINGS

The other half of the construction document is the plans or drawings, an integral component to convey the design intent of the systems. The drawings show locations, relationships, dimensions and detail. The design drawings are also organized and standardized. Drawings, or a drawing set, are typically organized by discipline (e.g., civil, structural, electrical, telecommunications), and then further by type of drawings (e.g., plan, elevation, section).

Note that drawings may also be prepared during construction by the contractor; these are called "shop drawings" and are used to show how equipment may be fabricated or installed (Fig. 13.3).

Manufacturers of integrated building system products need to understand how their products can be specified on a job. In addition, they must provide information regarding their products in a MasterFormat format rather than the traditional product data sheet. This indicates that the manufacturer

Drawing Series Identifier	Discipline
G	General
H	Hazardous Materials
V	Survey/Mapping
B	Geotechnical
W	Civil Works
C	Civil
L	Landscape
S	Structural
A	Architectural
I	Interiors
Q	Equipment
F	Fire Protection
P	Plumbing
D	Process
M	Mechanical
E	Electrical
T	Telecommunications
R	Resource
X	Other Disciplines
Z	Contractor/Shop Drawings
O	Operations

Figure 13.3 Standard naming convention for construction drawings.

understands the process, and may make it easier for the designer to specify the product. Contractors, especially those in IT who may be accustomed to purchase orders with an itemized list of equipment, must understand the contractual environment of new construction and the way their products will be identified, specified, and installed.

Understanding construction documents and the roles of the designer and contractor is important if the total benefits of integrated building systems are to be attained.

Design and Construction Data

For all of the data and information created and developed during the design and construction of a building very little of it is used in the operation and management of the building. It is not that the information and data would not be useful, but rather that much of the data is created in a static, legacy paper format. That makes it difficult if not impossible to use the data in facility operations, property management or business systems unless one decides to undertake the laborious task of manually entering data.

Lack of data and information severely hamper the operation and management of a building diminishing operational effectiveness, efficiencies and productivity. How serious are the inefficiencies and lack of interoperable data? In a recent report, the U.S. National Institute of Standards and Technology "estimated $15.8 billion as the annual cost burden due to inadequate interoperability in the capital facilities segment of the U.S. construction industry." A total of "$6.7 billion of the total was due to inefficiencies in the design and construction phases of the project delivery process." That was a conservative estimate.

The lack of useable data from the design and construction process impacts several "upstream" management systems. For facility management it involves systems for work orders, predictive maintenance, fleet management, inventory of materials and equipment and energy management. For business systems it includes asset management, purchasing, human resources and other aspects of an enterprise resource planning system.

At a minimum, one needs to integrate the building systems, a facility management system, and some business systems. Beyond that initial integration are other more global systems that have a potential requirement for integration such as property management and real-estate portfolio management.

None of that integration can occur without interoperable data and little data will be available if the owner has to manually enter data into the system after occupancy. Obviously, if information in electronic format were available when the asset was delivered, installed, or commissioned, the accuracy and comprehensiveness of the database would be much improved. Such a database foundation would ease system integration, thus resulting in the functionality, efficiencies, and cost advantages that integration can deliver.

Key data are created or provided in the following three progressions, or events of the design and construction process:

Construction Documents—These are crucial documents with legal standing, that lay out in a narrative and graphic format the owner's requirements. The specifications, done in a version of the CSI's MasterFormat, define the products, materials and expertise required. The drawings show quantities, graphic representations and relationships of the elements in the construction. These documents represent the owner's specific requirements as interpreted and refined by the architect and engineers.

Construction Process—Several documents are created during the construction by the contractor, subcontractor, suppliers, and manufacturers. These include shop drawings, coordination drawings, product data, samples, test reports, requests for information, construction change directives and plans.

Construction Closeout—These include record documents, addenda, field orders, change orders, spare parts and maintenance materials, warranties and commissioning documentation.

The vast majority of the documents created during the design and construction phases use word processing and CAD programs and are distributed via office paper and sheets of drawings. Much of this information is distributed via e-mail or posted on a FTP site but little of it is typically used or in a format to be used with the facility management or business systems of the owner.

If a building is being designed without the benefit of building information modeling (BIM) there are still approaches that can be taken to facilitate the gathering of electronic data. These approaches are not a replacement for BIM and will not provide the quality and quantity of information that BIM can but they will provide more information in an electronic format than the traditional paper plans and specifications. Summaries of such approaches follow:

- Have the contractors submit their product data electronically such as electronic copies of the manufacturer's specification sheets (in Adobe PDF) for each component delivered, installed, inventoried or made part of construction. These may not be in a database format but they can be stored in a data management application which may be a software module in some of the better facility management systems.

- Have the contractors provide a listing of the product data in electronic format, such as Microsoft Excel, and have it include item numbers, descriptions, item model numbers, order numbers, skew numbers, unit cost, preventative maintenance schedule, warranty and life cycle.

- Have the contractors supply the operations and maintenance manuals in Microsoft Word or Adobe PDF.

- Prior to construction develop a labeling scheme for all equipment and assets. Have the contractors use the scheme on all submitted drawings from shop drawings through as-built drawings. Naming conventions are especially important for campus environments and large real estate portfolios.

- Require that the administrative software for each building system utilize an open database, compliant with SQL and ODBC, such that it can share and retrieve information from other SQL and ODBC databases.

- Use a real-time location system (RTLS) to track the building's assets. The "legacy" method of tracking—bar codes—cannot be changed, needs a line of sight to be read, has a short life span and offers minimal security. Use of a wireless system (e.g., Wi-Fi, RFID, or Zigbee) offers greater data capacities, better security and may be read without line-of-sight or contact.

Building Information Model

Building information modeling (BIM) is the future of building design and construction. BIM is a 3-D, object-oriented, CAD approach for architects and engineers. While the number of architects and building designers using BIM is modest the number will continue to increase. One of the most valuable functions of BIM is its ability to improve the coordination between multiple design disciplines, thus reducing errors. BIM has the potential to respond to an owner's need for predictable costs, quality, and on-time delivery. (See Figure 13.4.)

The American Institute of Architects have called BIM a "model-based technology linked with a database of project information." It can store complete information about a building in a digital format including things like the quantities and properties of building components. It covers geospatial information and relationships regarding a building, and facilitates the digital exchange and interoperability of the data.

BIM uses the Industry Foundation Classes (IFC) for exchanging information about a building project among different CAD packages. XML, an Internet language, which allows raw data to be reliably shared over the Web, will also be used in BIM implementations. BIM has the potential to be the vehicle or depository for use by the design team, the contractors, and owner, with each party having the capability to add their own data and information to the model. The National BIM Standard (NBIMS) is being developed and major vendors have endorsed and supported the effort.

BIM has major benefits. One is the capability for BIM tools to detect "collisions," that is, design features that are incompatible and in conflict. For instance, assume that a water pipe designed by the mechanical engineer would be installed in a way that it goes through a steel beam designed by the structural engineer. BIM allows the design and construction teams to identify such collisions electronically rather than discover the collision at the construction site. The result is time savings and reduced construction change orders and related costs.

Probably more important is BIM's capability to provide the location, quantities, and properties of building components in product objects. Included in this information can be all details of components, such as manufacturer, model, warranty, preventive maintenance, and so on. This information is valuable in the operation and maintenance of the building.

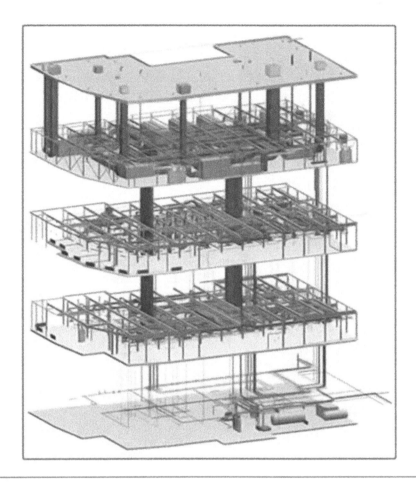

Figure 13.4   Typical building information model.

BIM is becoming more widely accepted for use in facility management. Starting in 2007, the U.S. General Services Administration (USGSA), under its National 3D-4D-BIM Program, requires spatial program information from BIMs for major projects receiving design funding. Four-dimensional (4D) models, which combine a 3D model with time, support the understanding of project phasing.

The American Institute of Architects (AIA) is modifying its contract documents to easily allow BIM, which is considered intellectual property, to make transfers from the architect to the facility manager, thus providing the facility manager with better data to manage a building.

The buildingSMART alliance, part of the U.S. National Institute of Building Sciences, provides useful tools to developers and users of BIM software and promotes the use of BIM. There are many important organizations that are a part of the buildingSMART alliance including the American Society of Heating, Refrigerating, and Air-Conditioning Engineers (ASHRAE).

The use of BIM may soon replace the Computer-Aided Facility Management (CAFM) process for facility managers. Typically the facility manager scans paper floor plans or imports electronic CAD files for use within the CAFM application. The electronic floor plans are then used to create "polylines" to define an area and identify room numbers to name that area.

For a typical commercial building, this process can take weeks. Instead, BIM files can be moved from the BIM creation software to facility management BIM software. The user can import the BIM file into software, which would include the room boundaries, room areas, room numbers, and space descriptions from the BIM. It would then perform the same functions as the typical CAFM software would but without all the lost time from the creation of "polylines."

In the not too distant future design and construction projects will require an information manager. This person or team will set the requirements for data management for the owner's project team, the design team, and construction contractors; manage the "supply chain" of data from design to construction to operations; and manage the integration of the data from the building and building systems into the owner's facility management and business systems. The drivers are economics, technology, increased functionality, and the overall value proposition.

Existing Buildings

With increased focus and higher visibility on new construction we tend to short-change the marketplace for making existing buildings smart and green. With existing properties making up 99% of building stock there are no reasons why existing buildings cannot be as smart or as green as new construction. In this section selected implementation issues in existing buildings along with a strategy for improving the performance of existing building stock are discussed.

Unlike new construction existing buildings come with existing building-technology systems and unique implementation issues such as the following:

- A number of systems, especially automation systems, will probably be using proprietary or legacy network protocols that must be migrated to open protocols. Typically, this may mean the use of gateways or some middleware to translate protocols.

- There is a lack of cable pathways in existing buildings, resulting in greater use of wireless devices.

- There are life safety issues related to abandoned cables because the older cable may have a sheath that could contribute to the generation of fire and smoke during a fire. In the United States the National Electrical Code requires unused cable to be removed during renovations. This can increase project time and costs.

- There may be organizational issues involved with facility management, security, and IT that have been established as separate departments and may be less likely to cooperate with an integrated approach.

Despite the challenges the financial metrics of improving the performance of a building and adding appropriate technology amenities can be compelling. The investment in an existing building is returned in several ways: lower operating and energy costs, more cost effective tenant improvements, premium rents, higher asset valuation, and a positive effect on capital planning.

DEPLOYMENT

While many building owners find the concept of upgrading the performance of existing buildings persuasive and intuitive they struggle with moving from a concept to actual deployment. The sections that follow describe a strategy to successfully improve the performance of existing buildings.

Go through a Discovery Process

Generally the information available on the system's performance in existing buildings is sparse. For example, if the building is using a typical BAS terminal from one of the major manufacturers the system is likely providing more raw data than useable, actionable information. Despite the meager amount of data, one has to pull together as much information as possible and analyze it to gain insight into the building's performance and to identify trends, pain points, and opportunities. One needs to gather information about:

- What the capabilities and features of the existing systems are
- What similar properties have done
- What the energy usage and costs for the building are
- What has generated most of the building maintenance work orders

Then survey the tenants and any third-party contractors to find out what changes or upgrades they think would be of value. Finally, physically inspect all systems.

Benchmark the Building's Performance

The only way to determine the effectiveness of the building's system upgrades is to have a baseline. After the upgrades, you will want to conduct a "before" and "after" comparison to judge upgrade effectiveness. Benchmark the current energy usage, the energy cost, and the number and type of work orders. Conduct surveys of tenants or occupants regarding their current satisfaction of lighting, security, technology amenities, and so on.

Decide Whether to Seek Leadership in Energy and Environmental Design Certification

Recent studies have shown that leadership in energy and environmental design (LEED) certification has a positive financial effect on the value of a building. Thus, if a building owner is upgrading systems and wants added financial value, incorporating LEED certification as part of the strategy makes sense.

A critical part of LEED certification for existing buildings (which focuses on operations and maintenance) is system upgrades. Much of what is often proposed as a strategy for existing buildings can satisfy some of the LEED criteria. LEED certification for existing buildings also requires addressing issues other than system upgrades such as recycling, exterior maintenance programs, and cleaning or maintenance issues.

Prioritize and Fund the Effort

One should prioritize and sequence the system upgrades based on their potential financial return and technical analysis. If you have a portfolio of real estate, prioritize on a building-by-building basis and pick out a couple of buildings to use as pilot projects. Set a budget for the upgrades and commit funding for the project. If funding is an issue, consider the use of an energy services company (ESCO). ESCOs essentially enter into an energy performance contract.

Initially ESCOs may be paid a fee for energy audits and a feasibility study. The ESCO may then determine the cost of the energy system upgrades and the potential energy savings and put together a business case for funding the costs of the upgrades in return for a portion of the energy savings. Some ESCOs may even fund the non-energy technology upgrades as well, depending on the business case. Also examine tax incentives and rebates from governments and utilities to offset the capital cost of the upgrades.

Upgrade the HVAC and Lighting Controls First

You will want to ensure that the major energy-related systems are improved to perform optimally. Prior to changing out any of the mechanical or electrical system hardware you'll probably need to upgrade the control systems and the control system terminal application software. This will allow the owner to obtain the needed information about systems, usage, and issues prior to investing in the more expensive mechanical and electrical upgrades.

Some system issues may not be mechanical or electrical in nature but can involve sequences of operation, changed use of spaces or undocumented system changes. Getting the information through a new set of controls and operator software will give you insight into how to effectively and efficiently recommission the systems.

Recommission the HVAC System

With the information obtained from the new control system and physical inspection recommission the HVAC system. Make the hardware upgrades or replacements as needed. Studies have shown that recommissioned systems in commercial office space have an average payback of 9 months. In a hospital environment where energy usage may be 2.5 times that of commercial office space the payback can be measured in weeks.

Upgrade the Building Security, Energy-Related, and Technology Amenities

The decision as to what other systems need to be upgraded is based on the estimated financial return of the investment. That is, will it lower operating costs, provide a basis for increasing rents or lower energy costs? The security systems obviously need to provide adequate security but they also need to be integrated into the lighting and HVAC control. An access control system in particular can provide information (i.e., number of persons in an area, when people are in an area, etc.) that can be used to adjust and manage the HVAC and lighting systems.

To better manage energy within the building you will need to add sensor systems (CO_2 sensors, occupancy sensors, etc.), metering for all energy and utility systems, and a power management system if the building lacks the same. Finally, if your building is commercial office space, examine those technology amenities that tenants may perceive have value. The idea is to turn the "value" into higher lease rates and tenants who rent for longer periods. These types of systems may include digital signage, Wi-Fi throughout the building and a distributed antenna system (DAS) to improve cell coverage.

Upgrade Monitoring, Management, and Operation of the Systems

All system upgrades, recommissioning, and LEED certification are for naught if the right tools and personnel are not in place to monitor and manage the systems and keep the building's performance in an optimal state. This is where open systems are especially important so that building system data, facility management systems, and business or property management systems can be fully integrated to provide actionable information on the building's performance. The information needs of a facility manager, chief financial officer or chief executive officer are different, but all will need to be addressed in deploying the operational tools.

The tools may require middleware, open databases, and some customization. Operating the building is more than simply software integration and Web dashboards. The skill sets of existing personnel in facility management, security, and IT may need to be examined and the organizational silos among security, BAS, and IT may need to be removed. If the building owner is upgrading buildings in a portfolio of real estate holdings, a centralized building operations center for all the buildings will prove to be the most effective and efficient method of monitoring and managing each building as well as the entire enterprise.

The Economics of Smart Buildings

Overview

Buildings have long life cycles, typically between 25 and 40 years, depending on the type of building and the intent of the original construction. The life cycle cost of a building includes the initial costs of the facility (concept, design, financing, and construction), as well as the long-term operational costs of the building (Fig. 14.1).

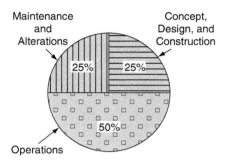

Figure 14.1 Life-cycle cost of a building.

Smart buildings can reduce both the construction cost of the technology systems as well as the overall building operations' costs. The cost savings from the smart building approach results in added value to the building as evidenced by lower capital (CAPEX) and operational (OPEX) expenses (Fig. 14.2).

Construction Costs

Construction costs for any building include the building technology systems that are at the heart of a smart building, namely telecommunications, building automation and life safety systems. There are construction costs for these systems whether the systems are installed separately or as integrated systems. The construction cost savings for integrated systems are primarily attributable to the efficiencies in cabling, cable pathways, labor, project management and system management hardware and software (Fig. 14.3).

When the technology systems are installed separately, the project involves more contractors and subcontractors. Each system's contractor is installing a system requiring cabling, cable pathways, equipment space, power, air conditioning, servers and management consoles. Each mobilizes a workforce. Each contractor must be managed, monitored and coordinated by the general contractor. This results in tremendous overhead costs (Fig. 14.4).

The smart building consolidates the cabling infrastructure and the number of contractors involved in construction. In an integrated system, the infrastructure cost, consisting of both cable and cable pathways, has a considerably lower cost for labor. Project management and engineering costs are typically reduced as well. The net result is that the smart building approach is less costly than installing systems separately. Removing the redundancy saves capital construction monies. Some examples are described in the following sections.

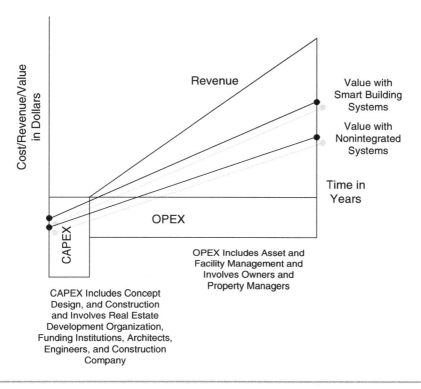

Figure 14.2   Distribution of building cost over its life and added value of smart building systems.

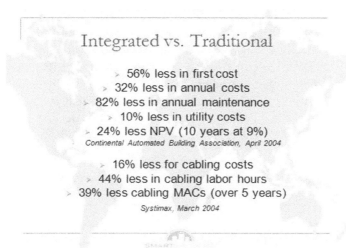

Figure 14.3   Two companies' Integrated versus Traditional cost differences.

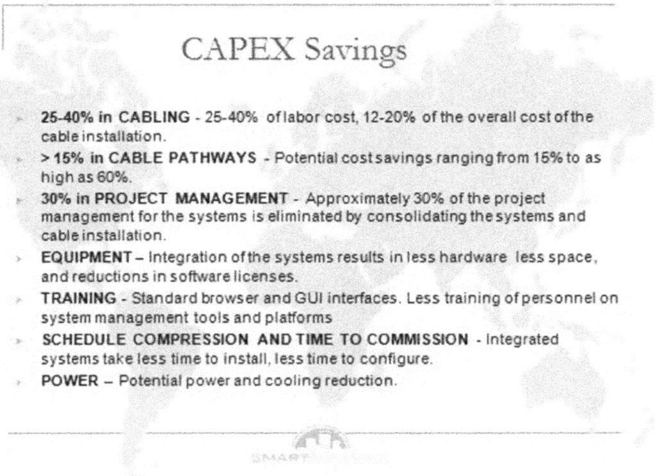

CAPEX Savings

- **25-40% in CABLING** - 25-40% of labor cost, 12-20% of the overall cost of the cable installation.
- **>15% in CABLE PATHWAYS** - Potential cost savings ranging from 15% to as high as 60%.
- **30% in PROJECT MANAGEMENT** - Approximately 30% of the project management for the systems is eliminated by consolidating the systems and cable installation.
- **EQUIPMENT** – Integration of the systems results in less hardware less space, and reductions in software licenses.
- **TRAINING** - Standard browser and GUI interfaces. Less training of personnel on system management tools and platforms
- **SCHEDULE COMPRESSION AND TIME TO COMMISSION** - Integrated systems take less time to install, less time to configure.
- **POWER** – Potential power and cooling reduction.

Figure 14.4   CAPEX Savings overhead costs.

CABLING

How many times during the construction of a building is Contractor A running a cable down a hallway on a Monday and Contractor B is running a cable down the same hallway on Tuesday? Given the fact that about 50% of the cost of cable installation is labor there will be efficiencies and cost savings if a single contractor installs both cables down the hallway. You cannot completely eliminate the labor cost of putting in the second cable, primarily because it still needs to be terminated, but depending on the cable density and type of building, 25 to 40% of labor cost is saved. That's 12 to 20% of the overall cost of the cable installation that can be saved when the cable infrastructure is consolidated and installed by a single contractor. Assuming that 30% of the cost of installing each of the building technology systems is related to cable that translates to a 4 to 6% saving on all systems.

CABLE PATHWAYS

If you consolidate the cable installation you can also consolidate the cable pathways. Pathways are the J-hooks, cable tray and conduit needed to install the cable. Costs for the cable pathways are part of the overall construction budget although they may be carried in the portion related to the electrical work. That is because the electrical contractor typically installs conduit and cable tray. Industry analysis has shown that the material cost of pathways is about the same in an integrated approach; but like the cabling, consolidation drastically reduces installation labor costs.

Additional efficiencies can be realized by addressing pathways for both low-voltage and high-voltage distribution in a coordinated approach. Potential cost savings are dependent on the type of space and the density of pathways, possibly ranging from 15% to as high as 60% in an open office environment.

Project Management

When the technology systems are installed separately the project has many more contractors or subcontractors involved. Each contractor mobilizes a workforce and has to be managed, monitored, and coordinated by the general contractor. Approximately 30% of the project management is eliminated by consolidating the systems and cable installation. Assuming that project management is 8% on top of the total cost for the installation of all systems, savings is on the order of 2 to 3% of total cost.

Equipment

Integration of the systems also involves consolidation of system servers. This consolidation results in less hardware, less space and reductions in ongoing software licenses. This may result in relatively small cost savings but it is still streamlining as a result of squeezing the redundancies and inefficiencies out of the legacy approach to system deployment.

Training

System integration also affects the number and functionality of the system administration workstations. Each system does not require its own dedicated workstation. Administration can be accomplished through standard browser and GUI interfaces. This allows a manager to access (as authorized) multiple systems through a similar interface and to do so locally or remotely, resulting in less training of personnel on system management tools and platforms, and less equipment. This may be a modest savings initially, but has a much larger impact over system life cycles. Quantifying the savings is project dependent.

Time to Commission

Time is money. Integrated systems not only take less time to install but also less time to configure. The use of shared standardized databases compliant with SQL and ODBC and tools such as XML and SOAP, allow for easier integration among building technology systems. They also facilitate the integration between building technology systems and the organization's business systems, such as purchasing, human resources, and so on. Less time to commission means less cost.

Power

POE (power-over-Ethernet) is the most undervalued technology for building construction and operation. Even though systems can be integrated without the use of POE the capital and operational savings from the use of POE is so compelling that it needs to be an integral part of the installation and operation of smart building systems.

For example, a card reader that is part of an access control system typically requires both a communication cable and local power. How much is saved if the communication cable can also provide the power? By eliminating the need for local power, upwards of $250 per outlet can be saved during construction and close to $750 after construction depending on the location.

The CAPEX savings related to a smart building approach is very dependent on building type, size, local construction market and so on. The greatest cost difference between integrated and separate systems is the long-term operational cost of the systems and the facility. Operational cost savings are related to the following factors.

- The standardized infrastructure allows for easier change implementations during the operational life of the facility for building automation system controls and devices, telephone systems, data networking, lighting, and other telecommunications and building systems.
- Increased building efficiency results in energy savings.
- Coordination and communication of systems in response to an emergency evacuation are improved. For example, integrated systems allow for the fire alarm, video surveillance, access control, HVAC, lighting control, and elevator systems to communicate during an emergency evacuation.
- Standardized management tools continue to reduce training costs.
- System information management is improved.
- Overall staff productivity is enhanced.

▓ Reduced costs result from the ability to competitively procure systems with open architectures and a generic structured cabling system.

▓ System management among many separate facilities through the use of the Internet or a private network can be consolidated.

▓ Integration with additional business systems, such as human resources and purchasing due to standardized databases, is easier.

For a building life cycle of 40 years operational costs may account for 50% of the total cost of the building. Operational costs are then equal to the total cost of construction, financing and renovating the building. A relatively modest savings in annual operational costs will garner significant savings over the building's life cycle.

Public and private organizations have weighed in on the cost savings in integrated smart building systems. For instance, a study conducted by the National Institute of Standards and Technology posed the question, "[D]oes it indicate that investment in CBS (cybernetic building systems) products and services by individual owners and operators will be cost effective?" The answer to that question is most certainly "yes."

This study concluded that "[f]or every dollar invested in 2003 approximately $4.50 is returned (i.e., a savings-to-investment ratio of 4.5). This equates to an adjusted internal rate of return of approximately 20% per year. . . ." The study conservatively estimated that integrated systems have annual energy cost savings of $0.16/sq. ft., annual maintenance savings of $0.15/sq. ft., annual savings for repair and replacement of $0.05/sq. ft., and annual savings related to "occupant productivity" of $0.39/sq. ft. (Information in this and the preceding paragraph is from R.E. Chapman, "How interoperability saves money," *ASHRAE Journal*, February 2001.)

Conclusions by the Continental Automated Building Association upon analyzing life-cycle costs follow. (The following list excerpted is from the Continental Automated Building Association publication, *Life Cycle Costing of Automation Controls for Intelligent and Integrated Facilities*, A White Paper for Task Force 3 of the Intelligent and Integrated Building Council, April 2004.)

▓ First costs for integrated systems (including management hardware and software, network upgrades, web services, and devices) were 56% less than nonintegrated systems.

▓ Annual costs for changes, alterations, and upgrades after a system's warranty period (including service contracts, additions and remodeling, software upgrades and reserves for systems replacement) in integrated systems were 32% less than nonintegrated systems.

■ Annual operating and maintenance costs (e.g., staff, training, IT support and management reporting) for an integrated system are 82% less than a nonintegrated system.

■ Integrated systems saved 10% in utility costs (including integrated lighting and HVAC, improved load factor, coordinated supply and demand strategies) as compared to energy costs for nonintegrated systems.

■ The NPV of the life-cycle costs of an integrated system (10 years with a discount rate of 9%) was 24% less than a nonintegrated system.

Systimax, a Commscope Company that manufactures and distributes the physical layer (cable, cable pathways, connectors, etc.) for smart building systems, prepared a cost model primarily related to initial cabling and infrastructure costs for technology systems and the ongoing costs related to the "churn rate" of moving, adding, and changing services. This model was based on a typical 100,000 sq. ft. five-story commercial office building using the same cable for all technology systems and a common pathway for all horizontal low- and high-voltage services. Systimax's conclusions that follow are from *Cost-Reducing Construction Techniques for New and Renovated Buildings/Cost Models*, White Paper Issue 2, March 2004.

■ The cable and cable pathway installation costs for integrated systems were 16% lower than nonintegrated systems and required 44% fewer labor hours.

■ The cable and cable pathway costs for addressing moves, additions, and changes over a five-year period for integrated systems were 39% less than nonintegrated systems.

Another study in 2005 by a consortium known as the Converged Building Technologies Group (CBTG) concluded "there is a strong case for the utilization of the integrated approach. There is clear evidence that there are both commercial and technical benefits. The system is effectively scalable over the life of the building, thus obviating large upgrade costs and minimizing ongoing OPEX."

The study specified a typical building: an eight-story office building with headquarters facilities, a total of 13,500 square meters of space with a ratio of 80:20 net to gross floor area and a maximum capacity of 1500 people. The model building had the building plant on the roof as well as in the basement and a central core 10 meters from the edge on three sides and 16 meters on the fourth side. It also included a BMS system with 2500 points, day 1 support for 1100 users plus peripheral devices on an IP network, 400 fire devices, 400 speakers, 42 cameras, 46 access control card readers, and 15 intruder alarm points.

The cost comparison of the integrated system approach versus the traditional approach to system installation showed the integrated approach to be

24.2% less expensive to install. This amounted to a 4.5% reduction in construction costs for the whole building. CBTG concluded that the construction cost savings were predominantly derived from labor savings. The study also examined operational costs using a life cycle cost analysis over 30 years. The study affirmed that the integrated system approach enabled faster maintenance and upgrade implementations resulting in a 37% operational cost saving.

Conclusion

The aforementioned studies and research were conducted at different times by different groups using different methodologies and assumptions. However, the broad conclusions reached by these independent studies are all similar: the integration of building technology systems increases the efficiency of building design and construction, and produce more functional building systems that improve the operational performance of the building while lowering costs.

Chapter 15

Audio Visual Systems

Audio visual systems are a complex topic as they encompass scores of different types of equipment and material, multiple technical standards, and rapidly changing technologies. Despite the complexity, and in many ways because of the complexity, smart building technology (standard structured cable system,

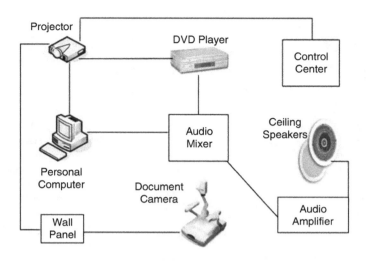

Figure 15.1   Typical classroom audio visual system.

Ethernet connections, and IP protocols) is taking hold in audio visual systems. This includes digitizing the traditional analog audio and video signals and more importantly, using the technology of data networks to control and manage audio visual systems.

There are a number of audio visual systems that can be deployed facility-wide; however, many audio visual systems are designed for the specific needs of certain rooms and spaces within a facility, such as meeting rooms or classrooms (Fig. 15.1). The core components of these audio visual systems follow:

- Audio and visual sources
- Processing and management
- Destinations (speakers and displays)
- System control

Audio and Video Sources

The sources of audio are generally from microphones, electronic instruments or programmed sources. The output from these audio sources is processed, adjusted, amplified, and fed to speakers.

Microphones convert acoustic energy (such as talking and singing) into electrical energy. Microphones come in a variety of types including handheld, wireless, lavalier or lapel clip-on microphones, headsets, and ceiling or table-mounted microphones. Electronic instruments, AM-FM tuners, audio CD players, cassette decks, MP3 players and personal computers are also audio sources.

The visual part of an audio visual system also has several sources:

- Cameras
- VCRs, DVDs, VTRs, DVTRs
- Cable television
- Satellite television
- Broadcast/cable/satellite
- Document cameras
- Personal computers
- Data networks

The wide range of video sources also means a wide range of video signal types and methods with which they connect to an audio visual system. Emerging video sources include electronic whiteboards and the Internet.

Speakers and instructors use personal computers as part of an audio visual system in meeting rooms, conference rooms and classrooms where presentations often rely on Microsoft PowerPoint software. The personal computer serves as a video input to an audio visual liquid crystal display (LCD) projector.

Document cameras have replaced the older, bulky overhead projectors. The document camera, which contains a video camera, is also typically connected to an LCD projector. Document cameras can handle a variety of materials including x-rays and even three-dimensional objects.

Interactive or electronic whiteboards can be used to display an electronic presentation but can also be used as an A/V source where everything written or displayed on a whiteboard can be recorded for electronic retrieval later. One application is in the classroom where an instructor utilizes the electronic whiteboard and the content on the whiteboard can subsequently be electronically retrieved by a student via a data network or the Internet (Fig. 15.2).

Audio and Video Processing

Audio signals may be processed in a multitude of ways with the most familiar being equalizers which are also used on many car and home stereo systems. An equalizer processes the audio signal by increasing or decreasing the low-, mid-, and high-level frequencies to provide the most pleasing audio sound possible. Brief descriptions of other common audio processors follow.

Figure 15.2 Interactive whiteboards.

Echo cancellation—This reduces the chance of sound output through a loudspeaker being reamplified into the sound input. Reamplification results in the well-known squealing feedback we have all heard when a microphone and a speaker are too close together.

Limiters—As the name implies, they limit the amplitude of the sound, restricting the volume to a predetermined setting.

Compressors—Compressors are like limiters but instead of just limiting the loudest sound they "compress" the audio signal to a certain volume range avoiding both the loudest and the quietest sounds.

Expanders—These simply expand or increase the volume of the audio signal to a predetermined sound level.

Gates—Gates block or eliminate audio sounds below a certain frequency level.

Automatic gain control—This automatically controls the volume of the signal at a specific frequency level.

Delay—Delay of an audio signal is typically used in large venues where natural sounds can be heard after processed sounds and delay is used to synchronize the two.

Digital signal processing—This is used with digital as opposed to analog audio and video signals.

Distribution amplifiers—Distribution amplifiers take a single audio input, amplify it, and distribute it to multiple audio outputs.

In many deployments of audio visual systems more than one sound source exists and an audio mixer is needed to combine or mix the multiple signals. For example, if there were three panelists speaking at a conference each with his or her own microphone, the mixer would combine and adjust the individual signals as needed. After the processing and combining the audio signal may be routed to a set of speakers, possibly using an amplifier in-between.

Mixers are used in recording and broadcast studios (mixing boards), but can also be part of a public address system and audio visual systems for large meeting rooms. Mixers have various features such as the ability to add effects to the sound and connections for a personal computer to provide enhanced equipment management.

Like audio signals, video signals also need to be processed, adjusted and "groomed." The processing of video signals is, however, significantly different than audio as the amount of information for video signals is magnitudes greater than that of audio. Video processing may involve amplifying or adjusting the timing, color, brightness, or contrast of the signal. Video processing equipment can include time base correctors to maintain the integrity of the signal and video processing amplifiers.

Audio and video signals both require devices to switch and route signals. An audio visual switcher is a "traffic-cop" type of device with the ability to connect a signal input to a single output connection. A matrix switcher or router is just multiple switches with the capability to switch multiple inputs to multiple outputs.

Video switches are primarily used in video production and distribution. For example, a live event such as a football game can be covered by several cameras and a video production switcher allows production personnel to easily switch between cameras. A matrix switcher may be used with a video distribution system within a building allowing for different video sources to be routed to different areas or rooms within a facility.

The amplification of audio and video signals increases the voltage and power for further distribution of the signal before delivery to its destination. Video distribution amplifiers typically take a signal output, amplify it to maintain signal quality, then distribute it to multiple outputs. A room having multiple displays where one video signal is to be distributed to all available displays will use a video distribution amplifier.

Speakers and Displays

Loudspeakers perform the opposite function of a microphone converting electrical energy back into acoustical energy.

TYPES OF SPEAKERS

Speakers vary in the range of audio frequencies reproduced and the dispersion pattern of the sound. Loudspeaker types include cones and compression drivers. Cones can be used for reproduction of the bass-, mid- and high-frequency spectrum, while compression drivers are used only for mid- and high-range frequencies. Since no single loudspeaker driver can accurately reproduce the entire range of frequencies several drivers may be utilized in a single speaker enclosure.

DISPLAYS

A video display device presents information visually. Video displays come in a wide variety of sizes and core technologies and are utilized in a broad range of environments. Selection of a display is based on the dimensions of the viewing area, lighting, and the type of content and materials to be displayed.

Display devices can be front or rear projected. Front projection is the preferred display in presentation environments (such as classrooms and meeting rooms) because it is less expensive than rear protection and requires much less space. The viewing image for the audience is projected onto a front screen which reflects the image.

Rear screen projection is preferred in bright, well-lighted environments. Essentially light from a rear screen projector is transmitted through the screen. The screen is made of material that transmits the light with relatively little distortion.

Rear and front screen projections can use LCD, cathode ray tube (CRT), or digital light projector (DLP) technology.

Major types of video displays include the following:

Plasma Screen

Plasma screens are basically a network of chemical compounds called "phosphors" contained between two thin layers of glass that, when excited by an electric pulse, produce colors, light, and a picture. The picture of a plasma screen is bright and rich with color. Screens can be as large as 80 inches. Plasma screens are thin, lightweight, have a wide viewing angle and offer good pictures under normal room lighting conditions.

Digital Light Processing

Television and projection digital light processing (DLP) technology is based on a semiconductor invented in 1987 by Texas Instruments. The basic technology in a DLP monitor consists of millions of microscopic mirrors that direct light toward or away from pixels. The pixels control the amount of light reflected off of a mirrored surface. DLP monitors have excellent color reproduction and contrast, and are lightweight. The monitors are deeper or thicker than plasma and LCD monitors, because DLP uses a lamp. DLP technology is used in many rear screen projectors.

Liquid Crystal Displays

LCDs use a fluorescent backlight to send light through liquid crystal molecules. LCD monitors apply voltage to the pixels to adjust the darkness of the pixels, thus preventing the backlight from showing through. Many LCDs double as computer displays. An LCD can be as large as 55 inches. LCDs are very thin, lightweight and have good color reproduction and sharpness.

Cathode Ray Tube

CRTs use a vacuum tube that produces images when a phosphorescent surface is excited. The dot pitch (DPI or dots per inch) of a monitor (the distance between the dots) is a measure of the quality of resolution. CRTs have a large dot pitch, which means a lower resolution quality, but CRTs produce a brighter picture. CRTs have excellent color and contrast but tend to be big and heavy.

Audio Visual Control Systems

A control system has to simultaneously manage all various components of the audio visual system. A presenter or user may want to lower a ceiling-mounted projector, start the DVD or perform other related tasks. Some elements of an audio visual presentation system that may need to be controlled appear in the list on the next page.

Local control of most audio visual systems may be a series of control buttons and switches located on a wall within the room, a wireless device controlling a projector, display or other component, or a touchscreen on a computer. Traditional system controls may include relays, remote controls, and proprietary manufacturer controls. More often the control system for audio visual systems consists of a personal computer, personal computer software, and a data network utilizing Ethernet and IP protocols.

Although digital audio and video content can be transmitted over a data network most of the evolution of audio visual systems to data networks has been in the administration and control of the components. While some components of an AV system may have direct network connections, such as projectors, many other components do not. Projectors, cameras, VHS/DVD players, motorized projection screens, lighting, and window shades or curtains are AV devices that can typically be controlled in an IP network.

Some of these components may connect to an "Ethernet interface" or bridge device which interfaces the various inputs and outputs from multiple components onto a standard Ethernet IP network. These systems add a network "on top" of an audio visual system to compensate for some component shortcomings in connecting to the network directly. In the following, example components are listed.

- Projectors
- Monitors
- Camera
- VCR
- DVD
- Lighting
- Screens
- Room access
- Curtains, drapes
- Volume
- Alarms
- HVAC

Room scheduling software can send out an invitation to the meeting and the program would schedule the meeting to the nearest location that can accommodate the invitees. The software can trigger the control system to turn on the AV components and to adjust the room temperature and lighting. Photoelectric light sensors would sense the amount of natural light and properly adjust the window shades.

The capability to administer an audio visual system through a standard smart building technology allows functionality previously not available. This includes remote control of the system over the network, enterprise-wide asset management, preventive maintenance, content delivery, component software upgrades, and more.

Remote monitoring allows a service technician to use a web browser to access and monitor all components of an audio visual system. The technician can turn a projector or other component on or off, run diagnostics on the equipment, centrally control all displays and the like. Remote communication

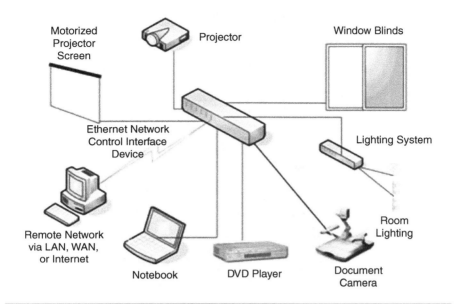

Motorized
Projector
Screen

Projector

Window Blinds

Ethernet Network
Control Interface
Device

Lighting System

Remote Network
via LAN, WAN,
or Internet

Room
Lighting

Notebook

DVD Player

Document
Camera

Figure 15.3   Audio visual system with Ethernet interface.

can be two-way. For example, technical personnel can be notified via email of an alarm or a maintenance event for a particular AV component, such as a burned-out projector bulb (Fig. 15.3).

Beyond system administration, management and monitoring through data networks, audio visual systems have increasingly incorporated the digital creation, sourcing, transport, display and storage of audio and visual signals. The wide variety of components and the types of cable interconnections means that adopting smart building technologies for audio visual systems is more complex than other technology systems.

Chapter 16

Network
Integration

Overview

Integrated building systems can truly save capital and operating building costs and have taken on a prominent role in green buildings. There are however different ways to "integrate" building systems, different concepts of what integration is, and a few misperceptions regarding integration.

Doi:10.1016/B978-1-85617-653-8.00016-8

A common framework for integrated building systems would make it easier to explain the approach and its value proposition to building owners, developers and architects. A framework could move the discussion to a common understanding. Clients would be better informed and designers and contractors may better understand what each is asking of the other. What follows are the methods and a framework for the integration of building technology systems.

Methods of Integration

System integration has different levels, different scopes of what is being integrated and different methods for how integration occurs. Not all integration is equal; that is, providing the same advantages and functionality. Integration can occur at physical, network, and application levels. Proprietary systems from a manufacturer may be integrated but not provide the flexibility and advantages of open systems. The following sections describe examples of system-integration methods and their attributes.

Hard-Wired

The most basic and oldest form of system integration is "hard-wired integration." A typical situation is two standalone building systems physically connected via an electrical "dry-contact" RS-232 or RS-422 connection. An example may be a fire alarm system connected to the HVAC control system or an access control system for secondary alarm annunciation and monitoring. These systems do not share any data but are simply connected to signal "off" and "on" conditions.

Proprietary/Bundled/Packaged Integration

This sounds like an oxymoron but a proprietary integration means that you use systems from one manufacturer that have been designed to work with each other. The downside to this approach is significant: The building owner is locked in to a single manufacturer for a complete system life cycle, there is no or little competition in procuring additional equipment and services, the building owner could be missing out on advances by other manufacturers, and no single manufacturer has the full suite of building technology systems. Some of the best examples of these "packaged" systems are in the residential marketplace, such as "home automation" packages for condominiums, apartments and single-family homes.

HANDSHAKE INTEGRATION

This involves two manufacturers of different building systems, or possibly facilities or business systems, agreeing to open their products to each other. The companies may have developed application programming interfaces (APIs) so their systems can communicate. This may work for two systems. However, managing all the APIs to get multiple proprietary building systems to talk to each other can be difficult.

OPEN STANDARDS INTEGRATION

Certainly using an industry standard protocol for the network layer of the building system is a required component in true integration. There are a couple of caveats to using open protocols however. One is that the number of protocols used to integrate the building technology systems needs to be minimized in order to maximize integration and provide an efficient operational environment. One approach is to select two major protocols for implementation (i.e., IP and BACnet/IP or IP and LonWorks), which in many cases will cover almost all needs of building systems.

The other, and maybe the more important caveat, is that the use of open protocols does not necessarily guarantee "openness," "interoperability," or "integration." Without certified or laboratory-tested products, protocols such as BACnet and LonWorks can be implemented in a way that may only be supportable by the original installer. The same situation exists with respect to the IP protocol. IP can carry proprietary data. IP by itself does not guarantee integration.

FRONT-END WORKSTATIONS

Another method, primarily used for BAS and security systems, is to have all systems networked to a single workstation. This was the approach taken by major BAS manufacturers beginning 10 to 15 years ago. This approach essentially provides a facility manager with a single standardized interface to manage and configure standalone systems. Each system may operate as a standalone system but the workstation provides a "unifying" tool and database for the systems. This is a somewhat proprietary integration, dependent on the manufacturer's application openness, and typically will address only a portion of the systems in a building.

The Framework for Referencing Integration

Integration of building system networks takes place at the physical, network and application levels of the networks. Integrated systems share resources. This sharing of resources underpins the financial metrics and improved functionality of integrated systems.

System integration means bringing the building systems together both physically and functionally. Physically refers to the cabling, equipment space and infrastructure support. It also touches on the common use of open protocols by the systems. Functionally systems integration addresses the capability of multiple systems to interoperate and thus provide functionality that cannot be provided by any one system. This is the theory that the "whole (integrated systems) is greater in functionality than the sum of the parts" (separate building technology systems).

There is a key differentiation between integrated and interfaced systems. Interfaced systems are essentially standalone systems that share data but continue to function as standalone systems. Integrated systems strive for a single database, considerably reducing the cost and support for synchronizing separate databases.

At the forefront of the evolution to open network standards is the International Standards Organization's (ISO) development of the Open System Interconnection (OSI) model, previously discussed in Chapter 2. The OSI model presents seven layers of network architecture (the flow of information within an open communications network), with each layer defined for a different portion of the communications link across the network. The OSI has withstood the test of time and this framework and some of its derivatives should serve as our reference point for integration.

Systems designers and contractors should frame the discussion of system integration using the ISO model and focus on the physical, data, network and application layers. It could very well add clarity and understanding to both industry and client discussions.

MIDDLEWARE

Middleware deals with the babble between building automation systems. Its objective is to bring communicative unity to disparate building technology systems. The benefits of middleware, as shown in Fig. 16.1, follow.

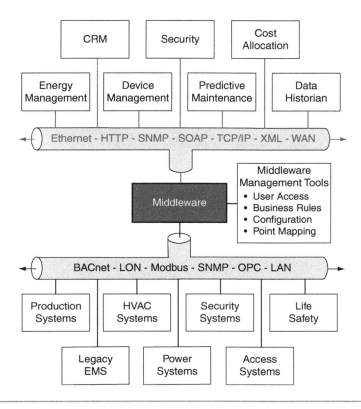

Figure 16.1 Middleware.

- Improving the functionality of existing systems. The bulk of systems the industry is dealing with are legacy propriety systems in existing buildings that have little or no integration and generate little or no meaningful information about the performance of the building. Middleware can leverage those existing investments.

- Allowing new construction to deploy the "best-of-breed" systems. Designers of new buildings can select the best individual systems and middleware can still be used to integrate these systems.

- Allowing for multiple user interfaces (web, mobile, etc.) and standard interfaces for all systems. This minimizes training on managing and controlling several different systems.

- Going wide, as in wide-area enterprise deployment. Middleware is at the heart of enterprise building operations centers.

- Providing for vertical as well as horizontal integration of systems. In brief, the information and data of the building systems should flow up to and synchronize with facility management and business systems. The integration with business systems is important as executives in their corner offices are now more interested in building performance. They are motivated to track energy use, analyze historical and real-time energy consumption, examine the types of work requests being generated, account for corporate assets, and so on. Middleware can unite the subsystems horizontally as well as coordinate facility management and business systems vertically.

■ Permitting shared information between systems. Middleware provides true software integration instead of hard-contact, point-to-point integration

The implementation of a middleware solution will normalize and standardize the data of building automation systems. In addition, these systems can extract and digest the data and control the systems such as set device values or set points. The word "middleware" may conjure, for some, the idea of software alone but implementation requires software and hardware. The hardware will involve subsystem controllers likely needed for the HVAC systems and some IT hardware, such as servers and workstations.

STRATEGY FOR DATA INTEGRATION PROCESSING

When applying data integration on a large project such as a college campus, airport, hospital, or even a mixed-use resort or commercial property, there are many details that must be evaluated. The first step should be listing the total number and types of systems that are going to be integrated at the building level and at the business level. These should be identified and organized per the 2004 CSI MasterFormat. A next step is to calculate how much data can be integrated from each system and subsystem, including physical devices and points, virtual values and calculated values.

The amount of data can add up quickly. When planning on using device data or values that are resident in field controllers or microprocessors embedded in an electrical or mechanical piece of equipment, exchanging that single piece of data or value between a business-level system based on the business rules of that system can be a challenge.

Careful attention must be paid to system and subsystem communications structures and limitations of baud rates and bus scans to be able to design a fully converged network to access the data at the IP network level as fast as possible. A classic example is a property management system (PMS) or facility management system (FMS) enterprise-level software platform that a building owner wants to use to automatically generate BTU/hour tenant bills for HVAC usage.

Where the calculation is performed in the BAS system, how often the values are calculated, the data format of the values, and the upload cycles between the systems are all details that need careful attention to ensure accurate customer billing. Data throughput, data scan rates, communication rates, and data exchange rules of system software, plus the total amount of values and data, are all part of designing the proper converged platform to meet the customer's demand for actionable information when it is needed.

Successfully Integrating Building Technology Systems

Many building owners and people in the industry find the concept of integrated building systems to be persuasive and intuitive. However, they struggle with moving from the concept to actual deployment. The steps to successful installation of integrated building systems are discussed in the following sections.

UNDERSTAND THE BUILDING OWNER'S BUSINESS

Integrating building systems is not a "one-size-fits-all" deal. There are different facility types and widely varying business objectives for private- and public-sector building owners. Integrated systems for a mixed-use project for a developer or owner are significantly different than for a 500-bed government hospital.

Since technology is simply used to further business objectives it is the business drivers of the owner that shape the approach for successfully integrating building technology systems. The discussion with an owner at this point has to do with how building technology systems can affect capital and operational costs, generate revenue, improve the experience or operation of building occupants, and enhance building operations or possibly differentiate the facility. Such a discussion is the programming or foundation on which to move forward.

GET EARLY PARTICIPATION IN PROJECT

The discussion with and buy-in of the owner has to be early in the project timeline. The later it is in the project schedule the less likely it is to be adopted, and if so, to be successful. Later in the process also means more disruptions. It is disruptive because some decisions have already been made and designs developed by the rest of the project team, specifically the architect and civil and mechanical, electrical, and plumbing (MEP) engineers. It is also disruptive given the budget implications that were not initially planned for. Chances for success depend on getting in early in the project.

Set Realistic Expectations with the Owner

"Under-promise" the owner. Steer clear of technology "futures" or "trends" and deal with what can be accomplished as of today. Identify for the owner any potential implementation issues and changes needed in the management of the building. Set an expectation with the owner that can be met or exceeded and the project will be successful.

Clearly Define Roles of "Traditional" Project Designers

The conventional way of designing building technology systems is to do it in "design silos." The legacy design and contracting methods do not work for integrated or converged systems. Success means an effort with the owner, the owner's facility or property manager, the architect, the mechanical engineer, the electrical engineer, the project team's consultants for IT, security and audio visual, the construction manager, and the contractors to adjust the design and installation process. Realize that some may resist doing things differently. Clearly identify who is designing and installing what and who will ultimately be responsible for the integrated system design and installation.

Detail the Scope—Clearly Identify Systems Involved

Specify what systems will be involved in the effort. Better yet, develop a matrix of all systems involved and determine what systems are to be integrated. Identify which systems will integrate on a physical level (cable, equipment rooms, etc.), a logical level (i.e., similar protocols), and a functional level. For example, the fire alarm will be integrated into access control and video surveillance, or the audio visual system will integrate with lighting controls and HVAC management. Move from the platitudes of "integrated systems" to the expected reality (Fig. 16.2).

Establish Systems' Technical Foundations and Operational Functions to Guide Design

Identify the common elements of the systems that are necessary for integration, can result in cost savings, or improve the operation of the building. These may be common cable types, a reduced set of communications protocols

Figure 16.2 Integration matrix.

Smart Building Integration Matrix©	Data Network	Structured Cable	Grounding System	VoIP	UPS System	Video Distribution System	Audio Visual System	Access Control System	Video Surveillance System	Intrusion Detections System	Wireless System	HVAC Management Control System	Electric Power Management Control System	Lighting Control System	Fire Alarm System	Elevator Systems Controls	Facility Management System	Integration of Business System
Data Network		•	•	•	•	•	•	•	•		•	•	•	•			•	•
Structured Cable	•			•	•	•	•	•	•			•	•	•	•	•	•	•
Grounding System	•			•		•	•	•	•	•	•	•	•	•	•	•	•	•
VoIP	•	•	•		•		•	•				•		•	•	•		
UPS System	•		•			•	•	•	•	•	•	•	•	•	•	•	•	•
Video Distribution System	•	•	•		•		•								•	•		
Audio Visual System	•		•	•	•	•						•		•				
Access Control System	•	•	•	•	•				•	•		•		•		•		•
Video Surveillance System	•	•	•		•			•		•					•	•		
Intrusion Detections System		•		•			•	•										
Wireless System	•	•	•	•	•		•			•								
HVAC Management Control System	•	•	•		•		•	•			•			•	•		•	•
Electric Power Management Control System	•		•		•												•	•
Programmable Lighting Control System	•	•	•	•	•		•	•	•	•		•			•		•	
Fire Alarm System		•	•	•	•			•	•			•		•		•		
Elevator Systems Controls		•	•	•	•			•	•			•		•	•			
Facility Management System	•							•	•	•	•	•	•	•	•	•		•
Integration of Business System	•								•								•	

allowed, open system databases, web-browser management tools, common equipment labeling schemes, and so on. Standardize, simplify, and find commonality among systems. Establish the technical foundations to guide the contractors.

BRIDGE GAPS AMONG PEOPLE INVOLVED IN FACILITY MANAGEMENT, LIFE SAFETY, AND INFORMATION TECHNOLOGY

Integrated systems affect the organizations that support and are responsible for those systems. The organizations' roles change slightly, the skills sets of the required personnel may change, and there may be budget implications.

Assist the owner in responding to these changes in order to identify improved operational efficiencies and cost effectiveness in the organizations.

PROVIDE COST ESTIMATES FOR PROJECT AND UPDATE COST ESTIMATES REGULARLY

After you provide an initial probable cost of the integrated systems, continue to update the cost based on market conditions or design changes. Minimize any surprises for the owner. Also, prioritize the items included in the installation in preparation for the "value engineering" discussion that will come at some point in the project.

UNDERSTAND AND ADHERE TO OVERALL PROJECT SCHEDULE AND ACTIVITIES' SEQUENCE

Large teams of people are involved in designing and constructing a building. The building systems are one small part of that effort, albeit critical to the operation and occupancy of the building. The design and installation of the building systems must follow an overall schedule and are dependent on several activities performed by others outside of the smaller integrated building systems group.

Here are a few things that will be specified and installed by others that the integrated building systems may depend on: space, conduit, cable trays, power, grounding, air conditioning, door hardware and furniture. Coordinate and sequence the schedule for integrated systems with the items needed for their proper installation.

DELIVER THE PLAN—DILIGENTLY MANAGE SYSTEMS' INSTALLATION AND OPERATION DETAILS

The best systems design plans and specifications are diminished if the systems are not properly installed or not installed to specifications. Manage the details. Continuously observe the installation. Inspect, inspect and inspect again, Work with others on the team to ensure success.

Chapter 17

Energy and Sustainability

Overview

Integrating a building's technology systems and constructing a sustainable or "green" building have much in common. Green buildings are about resource efficiency, life-cycle effects and building performance. Smart buildings, whose core is integrated building technology systems, are about construction and

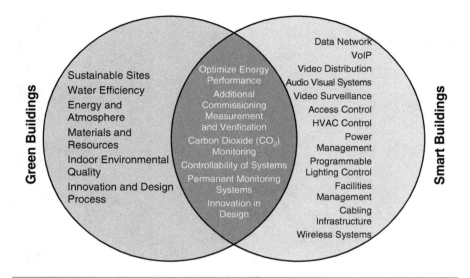

Figure 17.1   Commonality of smart and green buildings.

operational efficiencies and enhanced management and occupant functions (Fig. 17.1).

Part of what a smart building will deliver is energy control and energy cost savings beyond that of traditional systems installation due to the tighter control-system integration. Smart and green buildings deliver the financial and conservation benefits of energy management. Smart buildings are part of green buildings and greatly support and affect green building certification.

LEED

The Leadership in Energy and Environmental Design (LEED) certification is a rating system created by the U.S. Green Building Council (USGBC) to evaluate the energy and sustainability aspects of a building. LEED is an internationally recognized green building certification system created to encourage market transformation toward sustainable design. Formed in 1993, members realized that the building industry needed a system to define and measure "green buildings."

The USGBC began to research existing green building metrics and rating systems and the first LEED Pilot Project Program, also referred to as LEED Version 1.0, was launched in August 1998. LEED Version 2.0 was released in March 2000, with Version 2.1 following in 2002 and Version 2.2 in 2005.

On April 27, 2009, USGBC launched LEED Version 3.0. This constant updating of requirements for certification allows LEED to take advantage of new technologies and advancements in building science. In addition, it allows the USGBC to make adjustments to the certification, such as incorporating regional differences in sustainability and emphasizing operational performance rather than design intent.

Various LEED rating systems have been developed for different building types, sectors and project scopes. Included are LEED for Core and Shell, LEED for New Construction, LEED for Schools, LEED for Neighborhood Development, LEED for Retail, LEED for Healthcare, LEED for Homes, and LEED for Commercial Interiors. The Green Building Certification Institute (GBCI), established in 2008 by the USGBC as a separate entity, administers the certification programs related to green building practice such as LEED. The GBCI ensures that LEED buildings are constructed and operated as intended and its independence ensures that the results are unbiased.

LEED is credit-based, giving points for certain eco-friendly measures that are taken during the construction and use of a building. LEED is not rigidly structured and not every project must meet identical requirements to qualify. LEED is largely based on the following: energy savings, water efficiency, CO_2 emissions reduction, improved indoor environmental quality, improved building materials, use of resources and design innovation.

Credits in the LEED 2009 for New Construction and Major Renovations address seven topics:

- Sustainable sites (SS)
- Water efficiency (WE)
- Energy and atmosphere (EA)
- Materials and resources (MR)
- Indoor environmental quality (IEQ)
- Innovation in design (ID)
- Regional priority (RP)

In addition, there are eight prerequisites that each building must meet prior to any rating:

- Construction activity pollution prevention
- Water use reduction
- Fundamental commissioning of building energy systems

- Minimum energy performance
- Fundamental refrigerant management
- Storage and collection of recyclables
- Minimum level of indoor air quality
- Environmental tobacco smoke control

The four levels of LEED certification are based on points. All LEED rating systems have 100 base points; Innovation in Design (or Operations) and Regional Priority credits provide opportunities for up to 10 bonus points (Table 17.1).

Table 17.1 USGBC Ratings

Level	Points
Certified	40–49
Silver	50–59
Gold	60–69
Platinum	≥80

International Initiatives

Although it originated in the United States, many countries have expressed interest in the LEED program, including China and India. The World Bank estimates that by 2015 roughly half of the world's new building construction will take place in China and the majority of these projects are large, commercial office buildings of between 1 and 1.5 million square feet. In the past 20 years the Chinese government's top priority has been economic development so the construction industry has been constructing buildings as quickly and profitably as possible, taking little notice of environmental concerns. But that mentality is beginning to change.

A few progressive Chinese real estate developers are designing and building China's best green building projects without compromising economic growth. In 2003, the Century Prosper Center, a 1.6-million-square-foot twin office tower in Beijing's central business district, became the first large commercial project in China to be LEED certified. In 2005, USGBC President Rick Fedrizzi presented leadership awards to 10 Chinese real estate developers and government leaders for being the first in China to pursue LEED certification. By 2005, 10 large projects, with a total construction area of 5 million square feet, were LEED certified.

Individual developers are not the only ones contributing to this environmental push: certain cities are beginning to offer financial incentives to developers who use energy-efficient building technologies. The Ministry of Science and Technology, a branch of the Chinese government that coordinates science and technology activities throughout China, has become increasingly interested in LEED.

India, much like the United States, has its own green building organization, known as the Indian Green Building Council (IGBC). The IGBC, a part of the Confederation of Indian Industry, sponsors its own version of LEED, known as LEED-India. This rating system is very similar to the USGBC version of LEED; however, there are currently only two LEED certification types under the IGBC: LEED India for New Construction (LEED India NC), and LEED India for Core and Shell (LEED India CS). In addition to these two LEED certifications, the IGBC also sponsors two other certifications: the IGBC Green Homes Rating, and the IGBC Green Factory Certification, which was still under development as of this writing. Currently, the IGBC has certified 34 different buildings under their LEED-India program.

Canada has a green building organization as well, the Canada Green Building Council (CGBC), with its own version of LEED, LEED-Canada. According to the CGBC, the Canadian LEED certification is an adaptation of the USGBC LEED certification which is tailored specifically for Canadian climates, construction practices, and regulations, although for certain projects Canadian builders use the USGBC's version of LEED. The CGBC, which was established in December 2002, obtained an exclusive license in July 2003 from the USGBC to adapt the LEED rating system to Canadian climates and standards. At the time of this writing, the CGBC had certified 104 buildings through the LEED-Canada program, and 55 using the USGBC LEED program.

Other countries, including Mexico and Brazil, are currently working on local adaptations of the USGBC LEED program. Many countries that do not have LEED programs of their own are beginning to use the USGBC LEED certification. Australia, Germany, Sri Lanka, Israel, and the United Arab Emirates all have LEED certified projects. Some of the LEED-certified buildings were certified prior to development and establishment of national or local rating systems. For example, in the United Arab Emirates, both Abu Dhabi and Dubai recently instituted their own rating systems. In addition to the LEED certification, over 18 countries have green building certifications of their own.

Smart Buildings and Green Buildings

Integrated building technology systems align and support a green building approach. More specifically, a smart building approach facilitates meeting or exceeding the technical requirements of the credits and points of the LEED rating system. A few possibilities are discussed in the following sections.

WATER-EFFICIENT LANDSCAPING (2 TO 4 POINTS)

The intent of this credit is to limit or eliminate the use of water for landscape irrigation. Effective and efficient watering is left to irrigation control systems. One of several potential technologies and strategies for addressing this credit is to use climate-based controllers where irrigation is required.

Typically these are systems that have a preset schedule and sensors which can adjust the watering schedule as needed. These sensors may be moisture sensors, flow sensors, rain shut-offs and "evapotranspiration" controllers (devices that measure the evaporation of water into the air and the loss of water by the plants, used in precisely calculating the specific water needs of the plants). The result is watering the right areas with the right amount of water and avoiding over- or under-watering and runoff.

The irrigation systems usually have a central controller connected to a system administration terminal. These central controllers have evolved from motorized valves to electromechanical clocks to microprocessor-based controllers. The system administrators will use maps of the landscaped areas (developed using a GPS or the landscape architect's CAD drawings) to identify the locations of sprinklers and sensors. Some irrigation systems are fairly sophisticated with the system controller able to download current weather predictions and adjust the system as needed. Products are also available for organizations with a real estate portfolio to allow a central operations center to monitor each building's system and aggregate data for enterprise water use.

FUNDAMENTAL COMMISSIONING OF BUILDING ENERGY SYSTEMS (REQUIRED) AND ENHANCED COMMISSIONING (2 POINTS)

Fundamental building commissioning is a prerequisite for LEED certification. It involves a commission agent independent of the design team or the system contractors that is part of the design and construction process from the beginning. There are several reasons that commissioning of building systems has never been as critical as it is today. At the top of the list is the concern for energy and resource usage. Suboptimal performance of HVAC, lighting controls, alternative energy systems, water systems, and power management systems translates into inefficient energy and resource usage and increased energy, resource, and maintenance costs.

While LEED focuses on energy and resources, other building technology systems—security systems, fire alarm systems, fire suppression systems, telecommunications systems, information technology systems, and so forth—will be integrated into the base energy control systems. Commissioning has a substantial impact on the management, operation, and maintenance of the facility. If commissioning can produce even small improvements in the efficiencies and effectiveness of operations the result will be significant life-cycle cost savings.

Many building technology systems that are installed in new construction still perform at suboptimal levels. Surprisingly, a U.S. Department of Energy–sponsored study found "[t]he need for commissioning in new construction is indicated by our observation that the number of deficiencies identified in new construction exceeds that for existing buildings by a factor of three." It sounds counterintuitive but it's a reflection of common and pervasive design flaws, construction mistakes, and faulty equipment.

The study found that commissioning for new construction yielded a median payback of 4.8 years with one-time nonenergy benefits of $1.24 square feet. Nonenergy benefits included extended equipment lifetime, improved thermal comfort, decreased service call incidence, improved indoor air quality, first-cost reductions, labor savings, improved productivity and safety, decreased change orders and warranty claims, and liability reduction.

Total commissioning integrates commissioning into the total life cycle of a building with commissioning activities occurring throughout the design, construction and operation of the facility. It covers all of the building technology systems. It involves integration of the systems as required and uses software programs rather than paper checklists to make the most effective use of the data and information created, gathered, and stored.

Total commissioning starts early in the process. A commissioning agent is brought on board by the owner during the "concept phase" of the facility and joins the team that is developing the facility's program. The agent is focused on how to document the owner's requirements and defining the performance criteria that eventually will be measured during the system commissioning. Input by the facility manager is important at this early stage as well. The facility manager's input and participation during the commissioning process will improve the commissioning agent's work and can provide more relevant information for building operations and maintenance.

The requirements from the facility program and conceptual phase form the basis for the design. During the design phase, the commissioning agent checks to ensure that system requirements and measurable criteria are incorporated into the design. In many cases, this goes beyond the building technology systems. Assume for example that an owner wants office space to be at 70°F with 28% relative humidity. The commissioning agent will examine the design of the HVAC and lighting control systems, but also window treatments, insulation, glazing, and so on.

The design phase will produce construction specifications which can address issues related to commissioning. In CSI's MasterFormat 2004 commissioning is addressed in several places. Division 1—General Requirements cover both life-cycle activities and the commissioning agent. In the life-cycle sections the requirements for system performance are specified and the responsibilities of not only the commissioning agent but also the contractors are itemized. These references focus on the facility substructure (foundation and basement), the shell (superstructure, exterior enclosure, and roofing), and interiors (interior construction, stairways, and interior finishes). In addition, Division 1 addresses requirements for facility operation and maintenance. Division 25—Integrated Automation, Division 27—Communications, and Division 28—Electronic Safety and Security, all address commissioning of respective systems.

During the construction process the commissioning agent monitors changes and how they affect the owner's requirements, essentially providing quality assurance and control. These changes may involve material or equipment substitutions by the contractors, shop drawings, change orders, contractors' request for information, directives, supplemental instructions, and so on. The agent is involved with equipment start-up and operational testing and system transition into the hands of the facility manager and plant operator. More importantly, the commissioning agent checks against the owner's requirements identified in the conceptual phase.

The primary purpose of total commissioning is to ensure that the completed facility operates as originally intended. The information gathered during the commissioning process assists the facility manager in operating and maintaining the facility's systems. Information the commissioning agent creates is critical when establishing an operations and management program for the facility. The agent can provide information on routine maintenance, test reports, load testing, start-up data, equipment life-cycle, operating adjustments, training requirements, spare parts, and so on.

MINIMUM ENERGY PERFORMANCE (REQUIRED) AND OPTIMUM ENERGY PERFORMANCE (1 TO 19 POINTS)

These two areas are related. One deals with establishing a minimum level of energy performance for the buildings and systems, while the other addresses

LEED credits or points for going beyond the minimum energy performance and increasing energy efficiency beyond the minimum requirement. Buildings can receive 1 to 19 points for increasing energy performance beyond the pre-requisite energy performance. Typically, a building simulation using the "energy-cost budget method" is used to demonstrate energy performance and points are earned based on percentage improvement above the baseline.

One of the potential strategies to obtain credits and points is to design the building systems and building envelope to maximize energy performance. This focus is obviously on HVAC, interior lighting systems and service water heating for domestic or space heating purposes (referred to as regulated or nonprocess systems), all of which fall under the umbrella of a smart building.

Other systems, including plug loads, are referred to as unregulated or "process" systems. Plug loads comprise everything that plugs into the electrical distribution system, such as PCs, displays, cameras, vending machines and copiers. Plug loads make up 9 to 25% of a typical building's electrical load depending on the building type and density of devices.

An IP network is able to provide power-over-Ethernet (POE) to a range of "plug-load" devices. POE not only supplies low-voltage rather than high-voltage power to these devices, but more importantly, provides the means to control power to the device. Central control of the POE devices allows for devices to be turned on or off based on a predetermined schedule, a sensor, or an event, such as occupant use of an access card. The result can be reduced consumption of power to devices, lower power usage and a greener building. In addition, POE reduces the use of materials, thus eliminating the need to provide a power cable to the device.

The default process energy cost is 25% of the total energy cost for the baseline building. If the building's process energy cost is less than 25% of the baseline building energy cost, the LEED submittal must include documentation substantiating that process energy inputs are appropriate. For the purpose of this analysis, process energy is considered to include, but is not limited to, office and general miscellaneous equipment, computers, elevators and escalators, kitchen cooking and refrigeration, laundry washing and drying, lighting exempt from the lighting power allowance (e.g., lighting integral to medical equipment), and other devices.

Process loads must be identical for both the baseline building performance rating and the proposed building performance rating, although project teams submit an "exceptional calculation method" (ANSI/ASHRAE/IESNA Standard 90.1-2007 G2.5) to document measures that reduce process loads. This documentation requires a list of the assumptions made for both the base and the proposed design and information supporting these assumptions.

MEASUREMENT AND VERIFICATION (3 POINTS)

LEED credits are awarded if the project team can demonstrate that energy and water consumption over time have been accounted for. This involves metering systems and management tools such as an energy management application that can track actual usage and cost, configure "what-if" scenarios, compare projected cost with actual costs, diagnose faults and failures, and so forth. Such an application can be standalone or part of a portfolio of facility management applications and should be considered a standard component needed to manage a smart building.

The framework for measurement and verification is the International Performance Measurement and Verification Protocol (IPMVP). IPMVP is a guidance document that addresses determining and documenting savings resulting from energy-efficiency projects. It provides a framework and four measurement and verification (M&V) options for how savings can be transparently, reliably and consistently determined in a manner that enables independent verification. These options include a partially measured retrofit, a building retrofit, and whole-facility and calibrated simulation, where energy systems are characterized through simulation or engineering analysis. LEED does not confine the use of IPMVP M&V to energy systems.

OUTDOOR AIR–DELIVERY MONITORING (1 POINT)

LEED credit is given for a monitoring system that can provide data on the ventilation of spaces that can be used to adjust the HVAC system. The result is improved indoor air quality and occupant comfort. Such a system will monitor air flow and carbon dioxide (CO_2) levels. The monitoring sensors can be designed based on activity levels and zones or space use, then integrated in the building automation system, thus becoming part of a smart building. The system is required to generate an alarm when the HVAC system varies more than 10% from its designed performance.

CONTROLLABILITY OF SYSTEMS—LIGHTING AND THERMAL COMFORT (1 POINT)

LEED credit is provided when the building affords individual occupants or specific groups in multioccupant spaces (conference rooms, classrooms, etc.) the capability of controlling the lighting, temperature, and ventilation of their space. This level of individual control, while still maintaining overall system management, is part of programmable lighting and HVAC control systems.

Many times this type of control is provided to occupants through touchscreens or other smart building systems such as VoIP telephones. The requirement is to provide individual lighting controls for at least 90% of the building occupants to enable adjustments to suit individual task needs and preferences. For thermal controls, the requirement is for 50% of the building occupants to be able to adjust temperature and ventilation to meet individual needs and preferences. (Note that operable windows may be used in lieu of controls for occupants within proximity of the operable window.)

THERMAL COMFORT—VERIFICATION (1 POINT IN ADDITION TO THERMAL DESIGN CREDIT)

The intent of this LEED credit is to provide for the assessment of occupant thermal comfort over time with one of the requirements being a permanent monitoring system. The objective is through additional monitoring and sensors the smart building systems collect more data on system and occupant use, which is then turned into actionable information to ensure thermal comfort. Such a system documents and validates building performance, monitors and maintains the thermal environment, and provides information on corrective action.

INNOVATION IN DESIGN (1 TO 5 POINTS)

LEED will grant points for innovative ideas not covered by the Green Building Ratings System or ideas that substantially exceed a LEED performance credit. The intent is to provide an opportunity to the design team to achieve exceptional performance.

There are two ways credits can be achieved. One is the "Innovation in Design," in which a new energy or sustainability strategy is proposed that is not addressed in current LEED strategies. This allows the design team to obtain points for thinking outside the box. Potential innovations where integrated smart building systems would facilitate the reduction of energy and the reuse and recycling of materials seem limited only by imagination and creativity.

The other pathway is "exemplary performance." A point can be earned for exemplary performance when the project has exceeded the credit requirements by 100% or has met the next incremental percentage threshold of an existing credit.

One possible scenario could involve cabling of the building. A converged and standardized cable infrastructure for a smart building would require less cable and fewer cable pathways, thus reducing the use of materials and saving energy. Innovations such as zone cabling further reduce the use of materials. Standardization of the cable also allows for greater potential to reuse the materials, thus saving energy. Extensive deployment of wireless technology within a building further reduces the need for cable and minimizes the use of materials.

High-performance buildings cannot be just green *or* smart, but must be both. Smart buildings make green buildings greener, and green buildings make smart buildings smarter. The result is enhanced performance and functionality.

Chapter 18

Case Studies

Case studies are useful for understanding the foundations and deployment of smart building systems. While research and analysis of smart building systems are important, some of the driving forces of smart buildings such as technology advancements, energy conservation and financial efficiencies are intuitive, there is nothing more persuasive than actual demonstration of deployment.

There is a tendency to assume that universal metrics for smart buildings should exist regarding energy savings and capital and operating costs or a possible standardized technology approach. However the wide variety of building types and sizes, owners, designers, sites and economic scenarios makes that impossible. Each building needs to be placed in the context of factors that influence it. Case studies are a way to ascertain real-world results on how building owners and managers took the smart building approach and applied it to their situation. The two case studies that follow are examples of smart building technologies being used in different environments.

Ave Maria University

In 2002, Tom Monaghan, founder of Domino's Pizza and chairman of the Ave Maria Foundation and the Barron Collier Companies, a diversified development company, became partners in a major development near Naples, Florida, the town of Ave Maria. The impetus for the development was Monaghan's desire to build a permanent campus for Ave Maria University which became the centerpiece of the development. The campus is situated on about 1,000 acres, with the initial building phase consisting of 500,000 square feet of facilities, serving nearly 500 students and 200 faculty and staff (Fig. 18.1).

While the campus prides itself on social interaction, collaboration, community and personal instruction, the technology available to the students, staff, administrators and managers of the campus is exceptional. Installed in the first phase of the construction of the university were twenty-three integrated building technology systems. The project has been referred to as a "clear landmark in the area of intelligent building management" and has won global awards for best use of automation (Fig. 18.2).

PROJECT SCOPE

The initial construction phase of the university involved 11 buildings:

- Campus library
- Oratory
- K-12 facility
- Undergraduate dormitories (4)
- Student activities center

Figure 18.1   Ave Maria University.

Figure 18.2   Ave Maria University and town center, under construction.

- Academic building
- Central plant
- Student recreation center

The essential technical foundations for the specification of the building technology systems were mutlifaceted:

- The systems had to utilize industry standards and an open architecture capable of both integrating and interfacing various systems, networking hardware and software from different manufacturers. This prevented the installation of proprietary mechanical, electrical, and communications systems.
- The systems had to maximize the use of a common structured cabling infrastructure.
- The number of network communications protocols had to be minimized with the preferred open protocols being TCP/IP and Lonworks.
- The systems had to be provided with full-featured administration and management systems, allowing for management both locally and via a web browser.

The building technology systems deployed follow:

- Campus backbone data network
- Building data networks
- VoIP telephone system
- Network operations center
- Access control system
- Video surveillance system
- Audio visual systems
- Programmable lighting control system
- Fire alarm system
- HVAC control system
- Energy management system
- Electric power management system
- Campus cable systems
- Facility cabling systems

- Integrated telecommunications rack system
- Vertical telecom-room UPS systems
- Campus wireless system
- Sound reinforcement systems
- RFID system
- Facility management system (FMS)
- Computerized maintenance management system (CMMS)

RESULTS

The cost savings related to the Ave Maria deployment were significant. It was based on integrating and consolidating systems, infrastructure and organizations where possible. Some of the efficiencies in construction and operation include the following:

- The university combined facility management and IT groups into a single operations center. This provided some efficiency in equipment and labor and saved $350,000 per year from the departments' combined budgets.
- The project used a single cabling contractor to wire all systems. Overall, the project saved approximately $1 million in cabling cost or about 30%.
- After the university opened, the energy management tools allowed the university to continuously reduce energy use and costs. Ave Maria now spends about $600,000 less on energy per year than when it first opened in 2007.

BEST PRACTICES AND LESSONS LEARNED

Owner Driven—Ave Maria University was a smart building project driven by the owner. Bryan Mehaffey, vice president of systems engineering for the university, understood the technology approach and its effects on construction and long-term operation of the campus. It was Ave Maria University that directed the design team and the construction management team to design and install integrated building technology systems.

Single Contractor—All systems, related cabling, and the management and administration systems were procured from a single contractor, a building systems or technology contractor. That contractor, while proficient in building automation systems, was not able to install all systems and brought in subcontractors to supplement the installation. The single contractor provided a single point of responsibility for the systems and their integration. From an organizational perspective all relevant contracts were "integrated" under the technology contractor. According to Mehaffey, "Picking one contractor to do this project saved an enormous amount of not only money, but something more costly, being time. That's because we were able to reach out to one partner, consolidate all of the project management, mobilization and overhead costs into one platform."

Adjustment of the Design Team—The design team used a "Division 17" construction document as the vehicle for procuring the systems. The HVAC control system, the lighting control system, the fire alarm system, and the power management system, all of which traditionally would have been procured from the contractors for Divisions 15 and 16, were procured in Division 17. A single design engineering company was responsible for the preparation of the Division 17 specifications and plans. A matrix of responsibilities was prepared to identify responsibilities for contractors.

Single Cabling Contractor—In a traditional installation of 23 building systems, each system contractor would install or subcontract for installation of system cabling. The bulk of the construction cost savings from integrated building systems is related to how this cable infrastructure is installed, which in the case of Ave Maria was through a single contractor that was a subcontractor to the technology contractor. Consolidating the installation of the cable, the cable pathways, and the space for equipment resulted in several efficiencies, including less labor being required, less project management due to less coordination between contractors, and a reduction in cable pathways. A single cable contractor was able to procure a larger volume of materials at a lower per unit price, also reducing cost.

Organizational Consolidation—Ave Maria recognized that information technology infrastructure was evolving and penetrating other building systems, also that facility management was taking on technical tools and technology as a means to manage and operate buildings. In addition, it was acknowledged that some of the services provided by IT and facility management were identical; each had to monitor and manage systems, respond to users and occupants, deal with university assets, provide reports, and so forth. The university's initiative to consolidate organizationally was based on such knowledge and the savings and efficiencies to be gained. In addition, staff members were provided tools that allowed them to view all system management software from their Blackberries and smart phones.

One Card and One Meta Directory—Each student as well as each staff and faculty member on campus has a smart card. The smart card provides both physical and network access. Access to buildings, such as student access to dormitory rooms, is performed through proximity smart cards. The card is also used as a debit card for purchases of books and food, as well as checking out materials from the library. Behind the scenes is a meta-directory database coordinating data and database updates between systems. This allows credential and identity management across systems, such as access control, human resources, student records, as well as identification and access to e-mail, the Internet, and intranet access. Such a database provides campus-wide management to student, staff, administrators, and visitors providing both improved physical and network security.

Ave Maria established a new paradigm for the design and integration of building and communications systems and provided the university with the opportunity to achieve unprecedented new capabilities while drastically reducing costs. It reflects the evolution of building systems to an IP network, financial advantages for building owners to integrate their systems, and the important role of building systems in controlling energy usage and costs.

State of Missouri

In 2005, the State of Missouri initiated a high-performance asset management strategy for their real estate portfolio.

OVERVIEW

The state owned or leased over $4 billion of real estate and was spending $300 million a year for acquisition, renovation and upkeep. While parts of the initiative addressed space planning and conditions assessment that could drive capital planning, the state also announced a plan to reduce building energy consumption by 15% by 2010.

The energy plan was motivated by constant increases in energy costs, operating costs, and deferred building maintenance backlogs. Achieving the goal required that the state's energy management, technology, and communications became a critical part of its overall asset management strategy. The implementation included utility billing, metering data and automated enterprise management. The energy management, technology and communications are the focus of this case study.

To execute the plan, the state utilized a performance contracting vehicle with an energy service company (ESCO). The ESCO analyzed the buildings, identified and designed an energy-efficient solution, installed the required elements and maintained the systems to ensure energy savings during a payback period. The project was guaranteed to save the state $9.5 million every year through reduced energy usage, process improvements in facility automation, monitoring, management, and more efficient real estate portfolio management.

The ESCO was at risk because earnings were based on performance. If the energy savings were less than that guaranteed, the ESCO was required to make up the difference. The state paid for the services through budgets for utilities, operations, and maintenance. The state then obtained guaranteed energy savings and paid for the upgrade to higher-performing buildings through operational budgets. The total cost for upgrading facilities and control and information systems was $24 million, including fees to the ESCO of $18.5 million.

PROJECT SCOPE

The state's real estate portfolio was approximately 32 million square feet in 1,000 buildings. Existing buildings came with baggage because they already had building technology systems installed. In Missouri many of the automation systems used proprietary or legacy network protocols that needed to be migrated to open protocols though the use of gateways or middleware to translate protocols. In addition, the information available on the system performance in existing buildings was sparse.

The only way to determine the eventual energy saved was to determine a baseline. After the upgrades it was necessary to conduct a "before" and "after" comparison to calculate the effectiveness of the improvements. The benchmarking included current energy usage and energy cost.

Once a baseline was established, the project focused on integration of existing and new systems, the design and deployment of a statewide communication network to gather data from each facility, and the development of higher levels of the information management system such as dashboards, analytics, human–machine interfaces (HMIs), and integration of other systems.

The objective was to be able to remotely and continuously monitor and manage the building systems and the operating conditions, provide fault detection and diagnostics, use the tools for initial system commissioning, transform data from building systems into actionable information, and integrate or interface this system with the utility bill payment system and metering data. Several management systems within the larger asset management project, such as conditions assessment, capital planning, and work order systems were integrated into the energy management, technology and communications project.

The ESCO for the State of Missouri, Johnson Controls, managed the program and projects. In addition, Johnson Controls facilitated the interconnection of building systems at each facility to the state's backbone communications network to allow for enterprise monitoring and system management. The ESCO brought other specialty contractors to their team to implement the project.

One company, Gridlogix, has since been acquired by Johnson Controls. Gridlogix provided the core software application that normalized all the building system data, gathered real-time data on an enterprise level, then presented the actionable information to Web-based dashboards for monitoring and management. The middleware features of their software allowed data from various existing buildings to be standardized. Talisen Technologies was another team partner which enabled the enterprise solution with deployment of IT hardware and network connectivity at each site. Finally, ISCO International implemented the wireless systems solutions for subsystems and components for all wireless connectivity at each site.

The tool that was developed and deployed had the following characteristics:

- Centralized energy-management reporting tools that tracked, recorded, and reported energy consumption with respect to time, then provided time-series, energy-use profiles by any combination of sites, facilities, and meters.
- Middleware that normalized and standardized data created in different formats by the building systems.
- Assisted with utility demand response allowing each of the facilities to optimize energy use and cost based on time-of-use utility rates.
- Recorded meter readings adjusted for weather conditions at the time of the readings.
- Features provided for users to relate facility data (square footage, type of use, etc.) with meter readings and produce analytical reports of the relationships, comparisons, and trends.
- Automated the process to be used to bill tenants for energy usage (Table 18.1).

EXAMPLES

The State of Missouri started the project in late 2006, with deployment in 2007. Two of the initial pilot buildings, the Wainwright Office Building and the Truman State Office Building are examples of project results.

Table 18.1 Financial Metrics

Number of square feet of state facilities addressed	16,336,715
Fees to ESCO	$18 million
One-time cost on a per-square-foot basis	$1.10
Guaranteed annual savings for energy per square foot	$0.17
Guaranteed annual savings for operations per square foot	$0.40
Simple payback for energy in years	6.5
Simple payback for operations in years	2.8
Total payback in years	**1.9**

Wainwright Office Building

The Wainwright Office Building, located in St. Louis, Missouri, was built in 1981, comprises 234,599 square feet, and is occupied by about 693 people. In the baseline year (before involvement of the ESCO), annual energy consumption for the Wainwright Office Building was 33 kilowatt hours per square foot at a cost of $620,301 per year. Two years after the ESCO the annual energy consumption had dropped to 19.7 kilowatt hours per square foot. Over the 2-year period, $250,000 in energy costs were realized, substantially more than the guaranteed energy savings (Figs. 18.3 and 18.4).

Truman State Office Building

The Truman State Office Building, located in Jefferson City, Missouri, comprises 775,000 square feet (see photo). In the baseline year (before involvement of the ESCO), annual energy consumption for the building was 26.9 kilowatt hours

Number of square feet of State facilities addressed	16,336,715
Fees to ESCO	$18,000,000
One time cost on a per sq. ft. basis	$1.10
Guaranteed Annual Savings for energy per sq. ft.	$0.17
Guaranteed Annual Savings for operations per sq. ft.	$0.40
Simple Payback for energy in years	6.5
Simple payback for operations in years	2.8
Total Payback in Years	**1.9**

Figure 18.3 Wainwright Office Building reduction in energy consumption after two years.

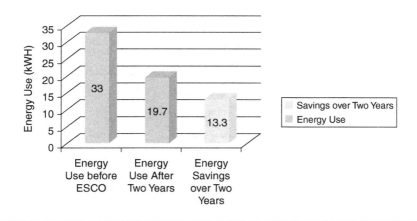

Figure 18.4 Wainwright Office Building reduction in energy costs after two years.

per square foot or $1.3 million in energy costs. In the first year after ESCO, annual energy consumption dropped to 21.2 kilowatt hours per square foot, or $986,220 in total energy costs, generating savings of $313,780. In the second year, annual energy consumption declined to 20 kilowatt hours per square foot and generated accumulated savings of $402,612 (Figs. 18.5 and 18.6).

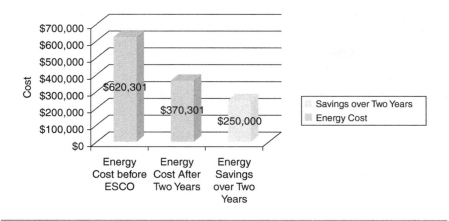

Figure 18.5 Truman State Office Building reduction in energy consumption after two years.

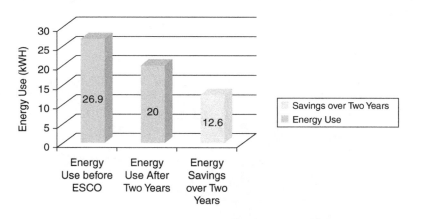

Figure 18.6 Truman State Office Building reduction in energy cost after two years.

RESULTS

Smart buildings provide actionable information about a building or space within a building to allow the building owner or occupant to more efficiently and effectively manage the building or space. This is what the State of Missouri accomplished. In Missouri, the building technology systems are integrated horizontally among all subsystems and vertically among facility management systems and other business systems. This integration allows information about the building operations to be utilized by multiple systems.

The project by the State of Missouri demonstrates the positive financial effects of integrated building systems and actionable information. The operational cost effects follow:

- Integrated and automated approach to energy use analysis and management and facility equipment maintenance management
- Enterprise-level information, presented in real time in dashboard views
- Sophisticated analytical tools that can correlate date from different systems and analyze a system, building or enterprise level
- Integration of stand-alone facility systems into a common information-based managed network
- Capability to react to changing consumption and weather variables
- Immediate fault detection and diagnostics, thus shortening the time between recommissioning of systems, resulting in enhanced system performance
- Centralized and decentralized (via the Web) access in a secure environment for energy and maintenance management applications, giving information visibility to technicians, managers, and executives
- Integration with other business systems such as asset management, space utilization, work order automation, conditions assessment, and capital planning
- Visibility into older facilities that prior to deployment was impossible
- Improved thermal comfort for building occupants, resulting in less work orders and increased occupant productivity

Appendix

Organizations and Associations

AEE—Association of Energy Engineers
4025 Pleasantdale Road, Suite 420
Atlanta, GA 30340
www.aeecenter.org

> AEE is a profession association involved in energy efficiency, utility deregulation, facility management, plant engineering, and environmental compliance, and also provides certification programs.

AFE—Association for Facilities Engineering
12801 Worldgate Drive, Suite 500
Herndon, VA 20170
www.afe.org

> AFE is a professional organization that provides education, certification, and technical information for plant and facility engineering, operations, and maintenance professionals.

AIA—The American Institute of Architects
1735 New York Avenue, NW
Washington, DC 20006-5292
www.aia.org

> AIA is the professional organization for architects in the United States.

AIIB—Asian Institute of Intelligent Buildings
City University of Hong Kong
Department of Building & Construction
Tat Chee Avenue, Kowloon
Hong Kong
www.aiib.net

> AIIB is an independent and academic institute whose aim is to develop Asia's definitions and standards for intelligent buildings and to act as an independent certification authority for intelligent buildings.

ANSI—American National Standards Institute
1819 L Street, NW, 6th floor
Washington, DC 20036
www.ansi.org

> ANSI is a private, nonprofit organization that administers and coordinates the U.S. voluntary standardization and conformity assessment system.

ASHRAE—The American Society of Heating, Refrigerating and Air-
  Conditioning Engineers
1791 Tullie Circle, NE
Atlanta, GA 30329
www.ashrae.org

> ASHRAE is an organization providing research, standards, publications and education on heating, refrigerating, and air conditioning.

ASIS—ASIS International (previously the American Society of Industrial
Security)
1625 Prince Street
Alexandria, VA 22314-2818
www.asisonline.org

> ASIS is an international professional organization involved with all aspects
> of security and provides certification for security professionals.

ASTM—ASTM International (previously the American Society for Testing
and Materials)
100 Barr Harbor Drive
PO Box C700
West Conshohocken, PA 19428-2959
www.astm.org

> ASTM International is one of the largest voluntary standards development orga-
> nizations for technical standards for materials, products, systems, and services.

BICSI—Building Industry Consultant Services International
8610 Hidden River Parkway
Tampa, FL 33637-1000
www.bicsi.org

> BICSI is an international telecommunications organization providing educa-
> tion and professional registration programs.

BOMA—Building Owners and Management Association
1101 15th St., NW, Suite 800
Washington, DC 20005
www.boma.org

> BOMA is an international organization of building owners, managers, devel-
> opers, facility managers, and other professionals, and provides education on
> office building development, leasing, building operating costs, energy con-
> sumption patterns, local and national building codes, legislation, occupancy
> statistics, and technological developments.

CABA—Continental Automated Buildings Association
1173 Cyrville Road, Suite 210
Ottawa, ON K1J 7S6, Canada
www.caba.org

> CABA is a nonprofit industry association that promotes advanced technolo-
> gies for the automation of homes and buildings in North America.

CANSA—Canadian Security Association
610 Alden Road, Suite 100
Markham, ON L3R 9Z1, Canada
www.canasa.org

> CANSA is a Canadian organization providing security education, government relations, marketing, communications, industry leader trade shows, and the latest industry information and news.

CSA International—Canadian Standards Association
178 Rexdale Boulevard
Toronto, ON M9W 1R3, Canada
www.csa-international.org

> CSA tests and certifies products to Canadian, U.S., and other nations' standards and issues the CSA Mark for qualified products.

CSC—Construction Specifications Canada
120 Carlton Street, Suite 312
Toronto, ON M5A 4K2, Canada
www.csc-dcc.ca

> CSC is a national association dedicated to the improvement of communication, contract documentation, and technical information in the construction industry, providing publications, education, professional development, and certification.

CSI—The Construction Specifications Institute
100 South Union Street, Suite 100
Alexandria, VA 22314
www.csinet.org

> CSI is a professional organization with a mission to advance the process of creating and sustaining the built environment. CSI provides information and education, and also produces the MasterFormat, the National CAD Standards, and the Project Resource Manual that are utilized for the design and construction of buildings.

EIA—Electronic Industries Alliance
2500 Wilson Boulevard
Arlington, VA 22201
www.eia.org

> EIA is an organization of electronic and high-tech associations and companies whose mission is promoting the market development and competitiveness of the U.S. high-tech industry through domestic and international policy efforts. EIA focuses on the areas of innovation and global competitiveness, international trade and market access, telecommunications and information technology, and cyber security.

ETSI—European Telecommunications Standards Institute
650 route des Lucioles
06921 Sophia-Antipolis Cedex, France
www.etsi.org

> ETSI is an independent, nonprofit organization responsible for standardization of information and communication technologies in Europe, including telecommunications, broadcasting, and related areas such as intelligent transportation and medical electronics.

FCC—Federal Communications Commission
445 12th Street, SW
Washington, DC 20554
www.fcc.gov

> The FCC is a U.S. government agency responsible for regulating interstate and international communications.

ICC—International Code Council
500 New Jersey Avenue, NW, 6th floor
Washington, DC 20001-2070
www.iccsafe.org

> ICC is a nonprofit organization dedicated to developing a single set of comprehensive and coordinated national model construction codes.

ICIA—InfoComm International Association
11242 Waples Mill Road, Suite 200
Fairfax, VA 22030
www.infocomm.org

> InfoComm is the international association of the professional audio-visual
> industries, providing education and certification for the audio-visual market.

IEEE
(previously the Institute of Electrical and Electronics Engineers, Inc.)
445 Hoes Lane
Piscataway, NJ 08854-4141
www.ieee.org

> IEEE is a nonprofit organization that promotes the engineering process of
> creating, developing, integrating, sharing, and applying knowledge about
> electrical, electronics, and information technologies and sciences. IEEE is a
> source of technical and professional information and standards.

IES—Illuminating Engineering Society
IES of North America
120 Wall Street, Floor 17
New York, NY 10005
www.iesna.org

> The IES has chapters throughout the world and is the recognized technical
> authority on illumination, providing information to designers, manufac-
> turers, engineers, and researchers.

IFMA—International Facility Management Association
1 East Greenway Plaza, Suite 1100
Houston, TX 77046-0194
www.ifma.org

> IFMA is a professional association for facility management, with members in
> 56 countries. IFMA certifies facility managers, conducts research, provides
> educational programs, and recognizes facility management degree and certif-
> icate programs.

ISA—Instrument Society of America
67 Alexander Drive
Research Triangle Park, NC 27709
www.isa.org

> ISA is a professional organization that develops standards, certifies industry professionals, and provides education and training, with a focus on instruments, systems, and industrial automation.

ISO—International Organization for Standardization
1, ch. de la Voie-Creuse, Case postale 56
CH-1211 Geneva 20, Switzerland
www.iso.org

> ISO is a nongovernmental organization consisting of the national standards institutes of 156 countries, with a central secretariat in Geneva, Switzerland, that coordinates the system. ISO is the world's largest developer of technical standards.

LCA—Lighting Controls Association
1300 North 17th Street, Suite 1847
Rosslyn, VA 22209
www.aboutlightingcontrols.org

> LCA is an adjunct of the National Electrical Manufacturers Association (NEMA). LCA provides education for the professional building design, construction, and management communities about the benefits and operation of automatic switching and dimming controls.

NFPA—National Fire Protection Association
1 Batterymarch Park
Quincy, MA 02169-7471
www.nfpa.org

> NFPA is an international organization providing and advocating consensus codes and standards, research, training, and education. NFPA's codes and standards influence every building in the United States, as well as many throughout the world. NFPA's code development process is accredited by the American National Standards Institute (ANSI).

NIBS—National Institute of Building Sciences
1090 Vermont Avenue, NW, Suite 700
Washington, DC 20005-4905
www.nibs.org

NIBS supports advances in building science and technology to improve the built environment, and has established a public/private partnership to enable findings on technical, building-related matters to be used effectively to improve government, commerce, and industry.

NSCA—National Systems Contractors Association
3950 River Ridge Drive NE
Cedar Rapids, IA 52402
www.nsca.org

NSCA is professional association representing the commercial electronic systems industry focusing on low-voltage systems.

OSCRE
799 Summit Drive
Santa Cruz, CA 95060
www.oscre.org

OSCRE is the organization responsible for the development and adoption of the Open Standards Consortium for Real Estate that drives the adoption of open e-business standards for the real estate industry.

SCTE—Society of Cable Telecommunications Engineers
140 Philips Road
Exton, PA 19341-1318
www.scte.org

SCTE is a professional association dedicated to cable television and telecommunications professionals. The organization provides professional development, information and standards. SCTE is accredited by the American National Standards Institute (ANSI).

SFPE—Society of Fire Protection Engineers
7315 Wisconsin Avenue, Suite 620E
Bethesda, MD 20814
www.sfpa.org

SFPE is a professional society representing fire protection engineering practitioners.

SIA—Security Industry Association
635 Slaters Lane, Suite 110
Alexandria, VA 22314-1108
www.siaonline.org

> SIA is an international association providing education, research, and technical standards for the security marketplace.

TIA—Telecommunications Industry Association
2500 Wilson Boulevard, Suite 300
Arlington, VA 22201-3834
www.tiaonline.org

> TIA represents providers of communications and information technology products and services and provides standards development and advocacy with governments. The TIA is ANSI accredited.

UL—Underwriters Laboratories
2600 N.W. Lake Road
Camas, WA 98607-8542
www.ul.com

> UL provides product-safety testing and certification within the United States.

Index

Note: Page numbers followed by '*f*' indicate figures.

CPSIA information can be obtained
at www.ICGtesting.com
Printed in the USA
BVHW032150290922
648368BV00003B/11

9 781856 176538